IN CHINA'S BORDER PROVINCES

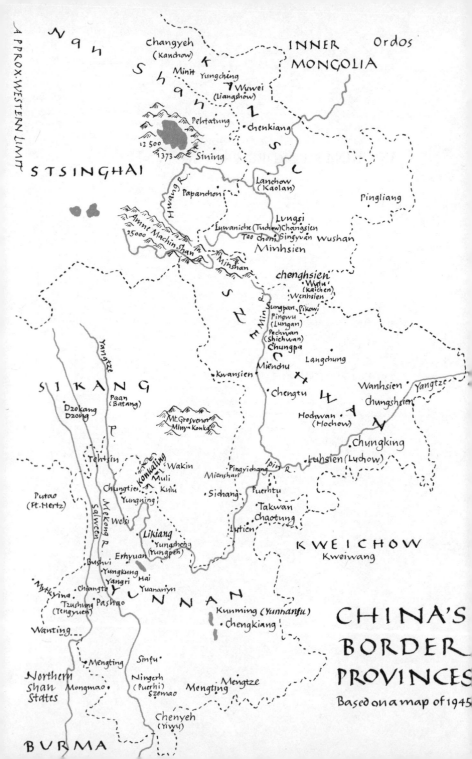

IN CHINA'S BORDER PROVINCES

The Turbulent Career of JOSEPH ROCK, Botanist-Explorer

by S. B. SUTTON

Illustrated with Photographs

HASTINGS HOUSE • PUBLISHERS *New York*

To Imre Halász,

with love

Library of Congress Cataloging in Publication Data

Sutton, Stephanne Barry.
  In China's border provinces.

  Bibliography: p.
  1. Rock, Joseph Francis Charles, 1884-1962.
2. China—Description and travel—1901-1948.
3. Botany—History. I. Title.
QK31.R62S9   915.1′03′40924 [B]   74-11238
ISBN 0-8038-3396-2

Published simultaneously in Canada by
Saunders of Toronto, Ltd., Don Mills, Ontario

Printed in the United States of America

CONTENTS

ILLUSTRATIONS

All illustrations by courtesy of the National Geographic Society, except for those specifically credited to the Arnold Arboretum of Harvard University

The Prince of Choni and Rock

The Tsungkuan of Yungning. Pavilion on his island of **Nyorophu**

I

JOURNEY TO THE
JADE DRAGON MOUNTAIN

On February 11, 1922, Joseph Francis Charles Rock, botanist-explorer on an assignment for the U. S. Department of Agriculture, reached the far western border of China at the head of an imposing caravan. The party consisted of Rock and two medical missionaries, six carriers, two guides, and one cook, with fifteen horses. For three months they had been inching their way into the mountainous wilderness from Siam, passing from one tribe to another through bandit-infested areas, and now had reached a small Chinese outpost known as Ban Ta Ngua, or the Crossing of the Buffalo, in the land of Yunnan.

"The moon rose shortly before our arrival at the wat." reported Rock in the diary in which he entered all his travels and moods. "The sun was just setting, and the hills in the distance

---

A note on Chinese geographic names: The rendering of Chinese place names into English poses problems. The modern *ping yin* system produces names which are unfamiliar to most readers, and the more common old Wade-Giles system is tiresomely pedantic. For the sake of simplicity, therefore, I have followed the lead of other modern authors and used the latter system minus the apostrophes or accents. Where a town name changed, I have indicated the alternate name in parentheses to assist a reader who might be using his own map. I beg the indulgence of scholars of China. S. B. S.

were purple. The sky was slightly hazy, and the full moon gently riding on the pale lilac haze, herself a pale silver disc, with the land masses clearly showing.

"A much-faded Chinese flag was implanted almost in the center of the road. To the left of it was a bamboo wooden shanty where the Chinese official, a small dirty fellow with a kindly smile, gave us a rather nice reception. Several Chinese were sitting about dressed in blue. They brought out stools for us and entered into a lively conversation. The official, whose name is Wa Kia-ki, like the beach in Honolulu, knew Mason (one of the missionaries) and without (our) asking for passports or anything we were shown the way to Chieng Law. He gave Mason some money to send down quinine. He had been on this border· for seven years.

"We are staying in a poor, miserable temple with two little alabaster buddhas and a small wooden one in the back. The wat is much crowded with Lus, a tribe of the Thai people. They were much interested in our lamp and the electric flashlight. They watched us eat, and when I put my plants away, they were crowding around me, coughing and belching all around and on top of me; I could stand it no longer. Many wanted medicines; one for an epileptic boy, the other for asthma, and many other ailments. We finally drove them off, telling them we were now going to bed."

China! The name was magic to Rock. Ever since his boyhood in Vienna, where he had been born nearly forty years before, he had dreamed about it, taught himself Chinese, and now was prepared to face reality. But the road he had taken was a hard one, and the dangers ahead were not limited to swarms of malaria-bearing mosquitos.

He already had faced situations that might have wrecked his caravan. For three months he had been traveling from the Burmese frontier to the Likiang snow range, heading east-north-east to Szemao and then bearing almost due north to Chingtung, Menghua (Weishan), and Tali, gradually climbing as he went. His route had taken him through areas in Siam inhabited by Miao tribesmen, and through Shan, Wa, and Kaw tribal villages in Burma, where the natives regarded his party with curiosity and fear.

As Rock approached China by the back door, taking the primitive road north through Kengtung, the threat of a dangerous confrontation with men or beasts was always present. Tigers roamed near the Burmese border and bandits lurked farther north, keeping Rock in a state of suspense. In the jungle villages at night, natives tossed green bamboo into the fire; it exploded with cracks louder than gunfire and was meant to scare off hungry tigers.

To keep bandits at a distance Rock had to depend on such soldiers as the local magistrates could provide, for they were supposed to protect foreigners. One day when Rock had only six soldiers armed with flintlock rifles he sighted a gang of fifteen robbers near a mountain forest. He reflected that all rifles look alike from a distance and "a bluff goes a long way," but to be safe had his caravan skirt the hill where the bandits had a strategic advantage.

Progress on the roads had been slow; the caravan was unwieldy. It moved about twenty miles a day, between fifty and eighty Chinese *li*, one *li* being about one-third of a mile. In the mornings Rock was out of his cot before dawn, the first man up, arousing his cook to prepare a heavy breakfast, for he was a hearty eater. Then, eager to start, he became irritated because the military escort had to be rounded up, carriers and muleteers had to get the gear secured to the pack animals. The soldiers straggled in last; Rock often correctly suspected they had sat up half the night gambling and smoking opium.

The day's advance had about four phases: a few miles of flat plain through cultivated fields; a long, steep ascent in mountain forests; a treacherous descent and a few more miles until the caravan reached a stopping place for the night. Rock, on horseback, liked to gallop ahead to take compass bearings and use the opportunity to collect exotic plants. Sometimes the midday halt came at a cluster of tea houses and food stalls, where the men could get some nourishment and the animals could be released to graze. Hopefully, a village was reached late in the afternoon.

The trail had climbed all the way from Burma. The caravan progressed from jungle to hills to mountains. Though backwards and isolated from the rest of China, western Yunnan was not true wilderness except in high mountain areas that defied life.

11

Elsewhere people lived wherever there was ground which could be cultivated. The road led Rock from one town to the next and through tiny hamlets. The land on either side of him showed signs of centuries of human occupation: rice, cane, vegetables and, farther north, wheat and opium. In many places the forests had been cut beyond salvation for firewood and building materials.

Rock passed coolies bent beneath their burdens, little girls carrying fifty or seventy-five pounds of salt strapped to their backs, mourners hauling coffins to a resting place. Near Hsinfu the caravan gave way to a wild horde of three or four hundred mounted Tibetans making their annual pilgrimage to Puerh and Szemao to purchase tea. South of Tali they met soldiers of the Eighteenth Regiment "looking as if they had been hatched from a Japanese cannonball," perfect imitations of the Japanese, even down to their hair styles. Like many other Sinophiles, Rock thought of Japan as a second rate copy of China with ill-considered Western inclinations; that the Chinese thought to copy the copy compounded the irony. The travelers along the trail were a motley assortment, and Rock learned gradually to distinguish variations among members of different tribes in physical appearance, dress, and language. The ethnic composition of his own party changed as he moved northward. His Thai carriers, whom he had hired in the jungle lowlands, balked at mountain travel; in their place he acquired Lolo tribesmen of whom he had a poor opinion, finding them filthy, afflicted with goiter, and prone to congenital idiocy.

Once in a town, Rock stopped to pay his compliments to the local magistrate, brandishing calling cards and credentials printed in Chinese. Over a cup of tea the magistrate advised him where to spend the night, the conditions along the road for the next stage, and how many soldiers he could provide, while adding sundry pieces of news about the town and politics and posing many questions. The quality of information Rock received varied from one magistrate to another. Sometimes he found the temple recommended to him for the night crowded with caskets of rotting corpses and crawling with flies and pigs, so he would look for lodgings in another temple or a schoolhouse. The carriers, muletiers, mules, and soldiers stayed at the so-called inns, but the conspiracy of filth, noise, opium smoke, and a remarkable assortment of vermin drove Rock wild, and nothing but desperation could lure him there.

The magistrate of one town, a heavy drinker eager to have Rock take his picture, looked blank when questioned about caravan routes and seemed ignorant of local geography. All magistrates, however, feared bandits and, because they were technically responsible for a foreigner's safety within their districts, calculated how many soldiers could be spared for an escort. Unlike the muletiers and their animals, which could be hired for several days, the soldiers seldom went more than one or two stages; therefore, the size of Rock's armed guard varied between six and twenty, depending on the number of soldiers in each town. Rock paid them according to pre-arranged terms. While soldiers were essential in bandit-ridden country, Rock remarked ruefully that often they gave "more trouble than actual help. They settle on a village like flies on a pie, and rarely pay for what they eat, but bully the farmers if the best is not forthcoming. . . . Much to their disgust, I made it a point, as long as any soldiers were with me, to make them pay for everything they took or ate."[1]

In Szemao, Puerh, and Chingtung, the larger towns, the caravan rested for a few days. The first—which one explorer, Ernest Wilson, had dubbed "the most God-forsaken place imaginable"[2]—had little to offer in the way of occidental diversions except the Customs officials and a lone Danish missionary lady who gave Rock clean lodgings. In Puerh, by contrast, Rock dined with the most powerful man in town, one General Liu. The ceremonies commenced in a courtyard spread with fresh pine boughs.

"Tea was served," the much-impressed explorer wrote later in his diary, "water chestnuts, small cakes, melon seeds, and pine seeds. We exchanged compliments and talked about the political conditions, Washington Conference, etc. Soup was served, and after some five minutes we were escorted into the dining room, the floor of which was also covered with pine branches. He served liquor first to myself, the guest of honor, and then to everyone else according to his rank. He motioned to each while pouring the wine, thereby indicating his seat. The dinner was excellent, consisting of sixteen courses and four dishes, which were placed first on the table. They were cold. Then dish by dish was brought, each helping himself to a large dish. There were whole stewed chickens, frogs' legs, and many other things that tasted very good but which I could not make out what they were. A white sweet liquor was served in small glasses, next an orange colored one in small silver cups. Sweet litchi stew finished the main dinner, but rice and some delicious soup were served in the end. After dinner we returned again to the sitting room where tea was served and nuts and fruits. The host, General Liu, was a very quiet, delightful gentleman who speaks very quietly in contrast to the very nervous and hysterical district magistrate."

Along with the general's hospitality, Puerh, a town of about 1,100 families, had other pleasant surprises. Rock visited the school where the boys learned English and teachers and students, citing the remittance of the American share of the Boxer indemnity in 1908, professed fervent friendship for the United States. Prevailed upon to "teach" a class in English, Rock talked about the United States, Hawaii, and his travels, then took photographs. That evening the teachers invited him to dinner in a temple. Rock remarked that the meal was good but not nearly as tasty as the General's.

Rock stopped to visit the salt mines between Puerh and Chingtung. The air in the mine shafts was foul, and the miners worked stark naked for pitiful wages. Rock felt revolted but refrained from extravagant emotions. He was learning, by repeated exposure, to observe without reacting to the horror of horrid conditions; he recorded his visit cooly.

Chingtung, unlike Puerh, did not receive him in a particularly friendly fashion. An American missionary lady of the Pentecostal variety, dressed in Chinese clothing, greeted him grimly and demanded to know if he were a Christian before she allowed him into the mission compound. Rock disliked the woman, a Mrs. Marston, on sight and was gratified to hear that the Chinese considered her rather queer. Though she had been in Chingtung for several years, she had not produced a single convert. Rock's soldiers were ailing, no one invited him to dinner, and crowds followed him around with resentful attitudes. He stayed for six days, keeping indoors as much as possible, attending to his plants and developing photographs. Mrs. Marston annoyed him, and he left her with relief.

Rock was not a hardy explorer, or one willing to compromise with the harsh conditions of a wilderness. He wanted his comfort and the amenities of western living. Only the most severe conditions could prevent him from setting up a linen-clothed table at dinnertime, complete with linen napkins, and commanding his cook to serve a proper Austrian meal prepared on a tiny brazier according to detailed recipes. He had not left his miserable household in Vienna to live like a pauper in the Orient. Fanatic about personal hygiene, he indulged in daily baths in an Abercrombie & Fitch folding bathtub and dressed fastidiously; he would not have dreamed of receiving, or being received by, even the most insignificant of chieftans wearing anything but a white shirt, necktie, and jacket. Rock's dedication to formalities meant that he had to transport more gear and, therefore, hire more carriers and horses than a less demanding traveler. He was not, however, entirely self-indulgent.

Aside from achieving agreeable comfort in unlikely places, Rock's life-style made a good impression on Orientals, and he was perfectly aware of this. Unlike many Westerners he understood, if only in part, the Eastern mystique of "face." That a wealthy European or American traveling in sloppy clothes, eating with his fingers from filthy tin plates, should lose face with his hired men, made perfect sense to Rock; that a coolie would not respect a foreigner who behaved as his equal also seemed logical. Confucius had written in his *Analects:* "When a prince's personal conduct is correct, his government is effective without the issuing of orders. If his personal conduct is not correct, he may issue orders but they will not be followed." Correct conduct indicated respect before superiors and boldness before persons of lower stations than oneself. Rock appreciated this aspect of the precise rules of face which dictated behavior and personal relations in the Orient. A dignitary would only be treated like a dignitary if he behaved accordingly; in Rock's own words, "You've got to make people believe you're someone of importance if you want to live in these wilds."

Rock entered China in 1922 as a skilled botanist and spent most of the next twenty-seven years of his life in the mountains along the Tibetan borderland. Until the mid-1930s he led expeditions and collected prodigious quantities of plant materials for various sponsors, including the National Geographic Society, the

U. S. Department of Agriculture, and the Arnold Arboretum of Harvard University. He was, however, a man of many talents and undisciplined curiosity. Without short-changing his scientific obligations he became progressively involved in studies of the tribespeople who inhabited the areas he explored. As he grew older and less able to withstand the rigors of the Chinese wilderness, he abandoned serious botanizing and engaged exclusively in cultural investigations. Having already published significant volumes on the botany of the Hawaiian Islands, where he had settled between his youth in Vienna and his adventures in China, Rock then produced an entire new body of scholarly work on the geography and ethnology of parts of western China. By the time he was 35 he was *the* authority on the flora of Hawaii; twenty years later he was also *the* authority on the Nakhi tribe of western Yunnan and, at least, *an* authority on that part of China which bordered Tibet. If Rock was eccentric, there was no cause for wonder; ordinary men do not decide to live in Chinese mountain villages.

For Rock, however, the Chinese wilderness was the fulfillment of an adolescent fantasy. In the early months of 1922 he was a mature man, 38 years old, an agricultural explorer for the Department of Agriculture. As he moved northward along the caravan trails of southeast Asia toward the Chinese border, he felt the anticipation of a child before Christmas. China lay, finally, ahead of him, within his grasp.

By April 20, Rock was near Tali, sometimes at elevations of 9,000 feet. On May 9 he got his first view of the Likiang snow range; the next night he camped amid blooming roses, oaks, apricots, and plums, in sight of the glittering Yulung shan, the Jade Dragon Mountain; on May 11 he led the caravan into Likiang. Of the dissonance between the illusion and the reality he would learn later.

❋ ❋ ❋

. . . .
rice grows and the land is invisible.
By the pomegranate water,
in the clear air
over Li Chiang

So wrote the poet Ezra Pound who, having exhausted Lt. Col. Waddell's Tibetan studies, turned to Rock's works for further inspiration; from the latter's dry prose, Pound extrapolated a fairyland.

Those who knew Likiang in reality rather than in imagination, however, saw it with more critical eyes.[3]

"What a queer contrary city is this Likiang!" complained Frank Kingdon Ward, the botanist, who had his headquarters there when Rock put in his appearance. "It should be clean, for sweet water babbles through the gutters down the steep cobbled streets; but it is dirty. It should be hot and muggy, for it is shut in by hills; but it is cold and draughty. It should be a great city and prosperous in trade, for all caravan roads meet here, from Tibet and the Marches, from Burma, and from all the rich cities in Western China; but it is squalid, poor, and dull. Daily the market square fills with petty traders from the surrounding hamlets, and no caravans go out laden with rich merchandise. It is a sorry place indeed; yet picturesque, with the snow-peaks to the north looming over it."

Kingdon Ward, like a number of other botanists, wanted imagination. Had he been less obsessed with plants and compass bearings, he might have found Likiang more attractive, though admittedly it was not everyone's idea of a good time. Still, another Western visitor, Peter Goullart, loved it, taking special note that the streets were "scrupulously clean."[4] Rock's offhand impression was not flattering: "A conglomeration of mud huts and a market place, which is to [the natives] a metropolis of marvellous splendor."[5]

Wary of the bustle of a commercial city of nearly 50,000 inhabitants, Rock chose to set up headquarters at 9,400 feet in Nguluko (Hsuehsungtsun), about twenty li from the town at the foot of the snow range and the highest village on the fertile Likiang plane. The local Nakhi tribesmen called it Nv-lv-k'ö, meaning "at the foot of the silver rocks."[*] The village, nestled at the base of the mountains and consisting of only about a hundred families, suited Rock's yearning for quiet while being conven-

_____

[*] Pronounced, according to Rock's explicit instructions, nasalized syllabic *v*, initial *l*, plus syllabic *v*-aspirated voiceless velar stop, fortis' half-high front viwel, rounded, as in German Sohne.

iently close to the large town. He rented a house with a tile roof, on a rocky street, and hired some Nakhi men as servants and assistants.

Rock's arrival, which ought to have cheered Kingdon Ward, failed to generate any enthusiasm in him. Having been "first" in Likiang in 1913, he viewed the newcomer as a poacher; that George Forrest had beaten them both by several years did not count with Kingdon Ward.[6] His reaction to Rock was a classic example of territorial jealousy. Western botanists in China staked out collecting areas the way gold miners do claims, but had no means of safeguarding what they considered their moral rights. Tales of Rock's stately entrance, flanked by soldiers, muletiers, and carriers, got back to Kingdon Ward and added to his anxieties. Rock evidently had plenty of money while he was hurting for cash. At the moment of Rock's arrival, Kingdon Ward happened to be out in the field, and Rock hired a couple of his erstwhile Nakhi servants; Kingdon Ward was absolutely livid when he learned what had happened and tried to get them back, but Rock paid better. Kingdon Ward complained bitterly that Rock had stolen his men. Predictably, after such a bad beginning, the two men never became friends. Rock was not the type to make amends or to try to justify his activities. No doubt the conflict of interest and personality partially explains Kingdon Ward's decision to concentrate his efforts thereafter in Assam and North Burma.

Rock, meanwhile, made this plain in the Yangtze loop, his base on and off for the next twenty-seven years, living sometimes in Nguluko, sometimes in Likiang proper. Until Peter Goullart arrived during World War II to set up experimental cooperatives on behalf of the Chiang Kai-shek government there were no other long-term foreign residents in the Likiang region save for a few missionaries, whom Rock was not disposed to count, and George Forrest, who came through on occasional botanical forays before his death in 1932; Rock first met him in May, 1922. Otherwise, Rock was pretty much on his own. The nearest Western settlement of any size was Yunnanfu (now Kunming), eighteen stages distant, where the French had railroad men and a consulate to look after the Hanoi-Yunnanfu railway, the British maintained a consulate to keep an eye on the

French, and an American consul watched British and French activities and dutifully reported his findings to Washington. Eighteen stages through bandit terrorized country was a long way to go in case of emergency, and Rock laid in year-long supplies of everything from medicines and tinned foods to chemicals for developing his photographs and blotters for drying plants.

From Likiang he went out into the field for weeks or months at a time, collecting, exploring, photographing, observing, and visiting practically every temple in his path. His first excursions in 1922 were to the Likiang snow range, where he made three short trips during the summer while training his Nakhi men to collect plant specimens and seeds. He taught one of them the methods of taxidermy and instructed another in the art of European cuisine. In September he took them south to Tengyueh, the Shweli Valley, and down into Burma in search of blight-resistant chestnut trees, then returned to Likiang in December. The Nakhis, who never before had been so far from home, rejoiced to be back.

Theoretically, Rock had completed his mission for the U. S. Department of Agriculture, but he indicated to them that the region was so rich in vegetation that he would like to stay on another year. The Office of Foreign Seed and Plant Introduction had already committed its exploring funds; rather than recall Rock, however, David Fairchild, the department's director, advised the National Geographic Society of Rock's whereabouts. The society agreed to pick up the funding of the expedition in Yunnan in February, 1923. Though Rock now operated under a new title, Explorer, with a new sponsor, the nature of his work did not change much. Under the aegis of the Geographic Society he amassed a grand total of 60,000 herbarium specimens, 1,600 birds and 60 mammals, which were publicly pronounced as being of "unusual interest and value," by Leohnard Stejneger, Head Curator of Biology at the U. S. National Museum, who received them from the Society. The USDA continued to receive seeds of interesting plants. The National Geographic Society got gorgeous photographs, maps, and eventually manuscripts, which the editors manipulated and converted into National Geographic-ese with an inattention to detail that infuriated Rock but which, from his self-imposed exile, he could not control.

THE NAHKI, A TRANSPLANTED TIBETAN PEOPLE

As Rock's, Kingdon Ward's, and Forrest's harvests proved, the river gorges and mountains of western China were sufficiently fertile to keep a man busy for a lifetime or at least as long as his legs held out. Without shirking his obligations to the natural sciences and a succession of high-powered sponsoring institutions —among them, after the USDA and the National Geographic Society, Harvard's Arnold Arboretum and Museum of Comparative Zoology, the U. S. National Museum, and the University of California at Berkeley—Rock began to take a scholarly look at his Nakhi hosts. Though not the first Westerner to stumble into their midst, he was one of the few to value their culture and to guess that it was only a matter of time before the Chinese engulfed and extinguished that culture with their own; he was the only one to attempt a full-scale study of the Nakhi.

Considerably sinicized, particularly in the town of Likiang where the Chinese influence was strongest, the Nakhi were descendents of Ch'iang Yüeh-hsi who originally immigrated from northeast Tibet. The Ch'iang tribe harrassed the Han dynasty by pushing into Chinese territory, but Han armies finally dispersed them into nine directions. The clan which Rock knew as the Nakhi moved south, first settling northeast of the Yangtze in western Yunnan and, about 25 A.D., crossing the river and taking possession of the land of the ancient P'u tribe within the Yangtze loop in the Likiang region. According to sacred Nakhi literature, their ancestors and heroes had hatched from eggs resulting from "copulations between mountains and lakes, pines and stones, *nagarajas* and human females," in the words of Peter Goullart's *Forgotten Kingdom*. A more profane history postpones their arrival in Yunnan until the reign of Hui Tsung of the Sung dynasty in the 12th century but confirms the Mongol origins by claiming the first ancestor to have been a bastard son of Kublai Khan named Yeh-Yeh. The Nakhi appeared to be related to the Moso, but history and mythology conspired to confuse the issue, and it was not clear whether the Nakhi absorbed, or were absorbed by, the Moso who preceded them in the Likiang region.[7]

The Chinese left the Nakhi to their own devices until the

Yuan dynasty (1280–1367) when the empire extended vague administration to western Yunnan, but successive emperors until K'ang-hsi decreed that Nakhi rulers control the region. During the first three centuries of Chinese influence, the Nakhi produced kings, heroes, and brave warriors, and enjoyed relative political integrity; simultaneously, however, they gradually embraced Chinese ways, effectively eroding their cultural identity—a situation which bothered purists and students of the culture more than it did the Nakhi. Finally, in 1723, the Manchus achieved complete jurisdiction in Likiang and withdrew administrative authority from the ruling Nakhi Mu family.

A people of intense racial pride, the Chinese patronized the "barbarians," a term they applied at random to all non-Chinese, and Rock unearthed a document produced by some 18th century Chinese Kipling, Kuan Hsueh-hsuan, the Manchu magistrate in Likiang, which asked winsomely, "Why did the Barbarians apply for naturalization in 1723? Why did they not remain in their peaceful barbarian state? . . . [Because] they had been attracted by the Imperial Benevolence as animals are attracted by sweet grass." Groaning self-righteously under their Yellow Man's Burden, the Chinese imposed their authority and tax collectors upon the Nakhi who, as it happened, inhabited an agriculturally desirable piece of land. As people who get pushed around and classified as barbarians are wont to do, the Nakhi despised their oppressors, the honorable Kuan's protests notwithstanding. Gradually, the declining Nakhi, unlike some of the other tribes in western China, resigned themselves to living in resentful obedience, exchanging their culture for that of their rulers.

By Rock's time Nakhi society was permeated with Chinese etiquette and forms; Nakhi men dressed like Chinese, but women retained their traditional costumes; only the priest-sorcerers or *dtombas*, chanting ancient ceremonies, could decipher the peculiar Nakhi pictographic script. Since the Chinese thought it beneath them to marry barbarians, the Nakhi were physically distinguishable, though, of course, there had been a certain amount of miscegenation. The Nakhi were typically darker complexioned than the Chinese, of average height or taller, well built, and handsome. "Other characteristics destroy any illusion that they have connections with the Chinese racial stock," Goullart observed. "Although the cheek-bones may be high, the face is essentially European in its contour The nose is long, well-shaped, and has a prominent ridge. Unlike the Chinese, a Nakhi gentleman could wear a pince-nez if he wanted to. The eyes are light brown and only in rare cases greenish; they are not almond-shaped, but wide and liquid." A few of Rock's Nakhi servants were over six feet in height—much taller than most Chinese—and very well built, towering over their retainer like gangland bodyguards. According to Goullart, who confessed to generalizations, the women, though forbidden to sit in the presence of men, dominated the society, did all the manual labor, masterminded the commercial life, and controled the pocketbooks in Likiang, thereby developing into robust, bold, shrewd, brainy creatures coveted as wives by less fortunate men of other tribes. The men, on the other hand, long stripped of their warrior status by the Manchus, had degenerated into loafers, drones, spongers, and opium smokers.

The Nakhi practiced a curious conglomerate of Bon—the pre-Buddhistic religion of Tibet—and tribal shamanism, accented with Chinese ingredients. To the dismay of earnest missionaries and to the undisguised glee of both Rock and Goullart they showed no proclivity for Christianity; they were just plain "inconvertible," according to Goullart; an earthy, superstitious, pleasure-loving people for whom Christian self-denial had no appeal. Some blame for the lack of conversions might be laid at the feet of the Christians, however, because they usually sent second-rate missionaries to the wild Chinese west, reserving the better ones for the coastal areas.

Rock romanticized the Nakhi, calling them Simple Children of Nature, Noble Savages, and crediting them with more innocence than they merited; he underestimated the mystique of his fair complexion and the power of his money. He lived like a potentate among them, surrounded by the precious artifacts accumulated on his expeditions; solid gold dinner plates, for example, a gift from the ruler of Muli; books, and his western gadgets including a battery-operated phonograph on which he played arias rendered by Caruso and Melba. Sometimes he went out for long periods in the mountains to collect plants and birds; every two years or so he decamped for America or Europe; then he returned to Likiang and his studies. All this seemed very odd indeed to the natives who had little experience with people who spent money and time on things without practical value. Still, he paid his servants well and was, on the whole, kind to them; high wages compensated for his occasional temper tantrums. Furthermore, unlike most Westerners the Nakhi encountered, Rock did not peddle Christianity. Since Likiang had no hospital, he treated the sick who came to him for help according to his medical supplies; he spoke both Chinese and the Nakhi dialect. He never fraternized with the natives or patronized the Likiang tea or wine shops—their version of the neighborhood bar—and he never invited the Nakhi to dine with him. He considered Goullart's chumminess with local people, in part indicated by the latter's mission to help them establish cooperatives, vulgar and, for his part, found it more expedient and comfortable to remain aloof. His lofty behavior, elegant living, evidence of wealth and generosity with same, and good manners of the Chinese style won respect for him in Likiang. Few Nakhi knew what he was up to, most thought him peculiar but, until it became fashionable to be anti-foreign, they found no reason either to mistrust or fear him.

Rock's scholarship began with a history of the tribe and a geographic study of the ancient kingdom, translated from Chinese manuscripts. From this preliminary but complex investigation he graduated to a study of the written Nakhi language. "What makes [the Nakhi] of paramount interest," he explained, "is the fact that of all the tribes in the region except the Lolo who are squatters in their immediate vicinity, they alone have a written language, in fact two types of writing: a pictographic, similar to Egyptian hieroglyphics, and a phonetic script. The latter is now little used, but an extensive literature is written in pictographs which is mostly of a religious nature but also gives an insight into their former mode of life when they were nomads in the far northwest . . ." Like an intelligence agent, he set out to crack the ancient pictographic code, something no other Westerner had tried.[8] Chinese scholars had toyed with translations with only partial success; Rock, with his passion for detail, eventually produced a bulky dictionary which not only translated but analyzed each Nakhi symbol historically, mythologically, and linquistically. His uncommon mixture of scholarly pursuits, which amazed and impressed academicians, came to him naturally, born of his scientific orientation, gift for languages, and fascination with the metaphysical. Habituated by taxonomy to scrutinizing microscopic characters and by field work to discovering large-scale vegetative patterns or generalities, he used the same methods in his ethnological research. What began as casual notes on Nakhi novelties for the *National Geographic* public evolved into an intense effort.

Rock had his first experience of Nakhi religious oddities in early 1923. "Today," he informed Dr. Fairchild, who expected him to be out collecting plants, "I witnessed a weird sight—the driving out of the evil spirits in a [Nakhi] house in this village. The wife of the owner of the house was sick or has been for some time. It was mostly due to fright and fear of the Tibetans which came so close to this village and besides her husband had to go and fight them. Well, there were three [Nakhi] wizards in full religious dress; in the courtyard they had erected what I would call a miniature garden, twigs of *Abies* and oak trees stuck in the ground surrounded by pine sticks dipped in yellow paint; on small, crude pine boards [Nakhi] gods were painted, and they were stuck about near the *Abies* twigs. At the end of this square garden was a table full of wheat seed, old eggs and dry peas, and all kinds of figures made of dough. Yellow, white and purple flags were stuck about with prayers written on them. They had a regular menagerie made of flour dough, snakes drinking out of cups, goats, sheep, etc. The priests then danced around this, one using brass cymbals, the other a gong which he struck with a long sword. One was beating a drum, and to all this humdrum foolishness the sick woman looked on . . . I felt the woman's pulse, and she had plainly fever. I gave her first a dose of castor oil which will do her more good than all the humdrum noise of the . . . priests."[9]

The wizards chanted from the ancient manuscripts which would become the bases of Rock's linguistic research. Picking his way through the maze of quaint pictographs, he employed a succession of *dtombas,* the Nakhi priest-sorcerers, witch doctors who, in the course of their professional duties—exorcising evil spirits, curing lumbago, invoking mountain gods—were the only people who used the script. Since the language and religion were intimately related, Rock needed to observe ceremonial proceedings, and it was not uncommon for him to rush half way across town to watch while a *dtomba* slaughtered chickens or pigs to rescue a soul from calamity. Rock's servants, alerted to his desires, kept a lookout for special occasions, sometimes running in from the street with the good news that so-and-so is dead or so-and-so is sick, or even so-and-so has committed suicide—an abnormally common occurrence among the Nakhi—and there is a ritual. Sixteen different funeral possibilities and over a hundred ceremonies in all varied the repetory. When he missed part of a ritual, was confused about its meaning, or wanted photographs in a better light, Rock asked a *dtomba* to repeat his performance another time for a fee, a practice which must have seemed strange to the Nakhi. Sensitive to worldly goods, *dtombas* were usually willing to oblige Rock, but their usefulness varied. Some feared that by revealing too much they would diminish their powers. In the 1940s Rock finally found a sorcerer who was intelligent, reliable, literate, and friendly; when he returned to Likiang after the war, however, the man had disappeared, and Rock had a terrible time finding a replacement.

As the years passed, Rock's age and running battle with intestinal disorders and other infirmities made expeditions more difficult. Always restless, he liked to get out into the mountains, and collecting plants and bird skins paid well, which the obscure academic products of his Nakhi research did not. His two volume Nakhi dictionary, which he did not live to see in print, cost him several thousand dollars. Yet he had a sense of urgency about completing his tribal studies which did not apply to botany; in fact, he produced *no* botanical papers related to China. It seemed more likely to him that someone else would carry on with the vegetation than with the ethnography; meanwhile, Nakhi culture risked being absorbed, once and for all, by the Chinese.

Every schoolchild learns that Champollion deciphered the Rosetta stone, but nobody bothers to find out what the Rosetta stone *said*. So it was with Rock. Whatever he achieved for the posterity of the Nakhi is the private property of a handful of scholars. The scholarly Harvard-Yenching Institute complains of piles of original Nakhi manuscripts which it acquired through

Rock but which no one there can read. Peter Goullart's glossy account of Likiang, meanwhile, has made it into paperback. Rock won his small fame on his expeditions, not because of the plants or birds or dictionary but in the pages of the *National Geographic,* where his hair-raising adventures along the Sino-Tibetan borderland and his encounters with exotic phenomena tantalized armchair imaginations.

II

THE HARD BEGINNINGS OF A YOUNG EXPLORER; FIRST BOTANICAL RECORDS IN HAWAII

Joseph Rock's China fantasy started in Vienna when he was a teenager. It had been a refuge from what he remembered, in later life, as a miserable childhood. He looked back upon it bitterly, always choosing to record unhappy, traumatic events. These somehow blotted out whatever childish pleasures may have come his way. Comparing himself with little David Copperfield he concluded that "his sufferings are analagous to mine as a boy . . . There is a little more tragedy in my life."

Rock was born in Vienna under the name of Josef Franz Karl, later Americanized to Joseph Francis Charles. His earliest and only fond memories were of his mother, Franciska Hofer Rock, a soft, warm woman of mixed Austrian and Hungarian descent. But she died in the summer of 1890 of an abdominal ailment; she was only 45, and little Joseph only 6. The child reacted naturally with a mixture of grief, bewilderment, and resentment, which his elders could not cope with. The funeral turned into a macabre affair. When Franciska lay in state, Joseph was required to place a flower in her lifeless hand, an experience that haunted him for the rest of his life. Then, precisely two weeks after Franciska's death her own mother died. The grandmother had been Joseph's single source of comfort.

Within these few weeks, the Rock family pattern altered. Joseph's older sister, Karolina, (Lina), became a surrogate mother to her brother at the age of 13, and dour Franz Seraph Rock dominated his son's life. Wretched with sorrow and trying to fill the role of both parents, he led the boy over and over again to weep beside his mother's grave, idealizing her memory. Finding no outlet for his agony and confusion, Joseph began to withdraw into his imagination.

Franz Rock looked a little bit like the vengeful God of the Old Testament, with a great white beard, white hair, and an intense gaze. Given to violent tempers, he did not tolerate contradictions or stupid questions, and inspired more fear than love in his children. A man of no particular intellectual accomplishment he apprenticed as a baker and, during Joseph's childhood, served as steward in the house of a Polish count who wintered in Vienna. Franz Rock was at once superstitious and deeply religious; he was, by most standards, a religious fanatic. Rigorous observance of the laws of the Catholic church and faithful attendance at services failed to satiate his cravings, so he devised an altar at home and, before it, he and the children play-acted the holy mass. He worked feverishly for the redemption of his soul. His outstanding sin was to have fathered an illegitimate child before his marriage to Franciska. Franz had offered to marry the mother, but she refused. Franz did not let the matter drop; he gave financial assistance, and made the child, Anton Maller, welcome in the Rock household.

The Rock quarters downstairs in Count Potocki's *Winterpalais* at No. 12 Schottenring became the scene of quarrels and tears. Joseph, who was sent part of each summer to visit his mother's relatives in Hungary, tried to run away to live with them when he was about 8. He succeeded in getting outside of Vienna before a stranger, pretending to help him, snatched his little satchel of clothing and food he had taken from the kitchen cupboard. Frightened and confused Joseph returned home, faced his father's wrath, and retreated further within himself. Lina grew jealous of her brother, believing him to be spoiled while she had to do all the work. Joseph returned her sentiment in kind, resenting her authority, and also became jealous of Anton because he suspected his father of preferring the bastard son. Adding to the snarl, Lina developed an indiscriminate adoration of children and collected waifs, thus further dividing her affections and nourishing Joseph's sense of neglect. One of these strays, Rosa Wetzel, joined the household when Joseph was 11 and stayed after he had left Vienna.

School only bored him. He learned quickly and remembered well; but while duller children recited their lessons, his attention wandered. At the Benedictine primary school he found a temporary friend in one teacher, Pater Alois, but the friendship did not offset his anxieties. He found alternatives to school more satisfying than the stodgy classroom where imagination was not encouraged and discipline was rigid. He played hooky, visited his mother's grave, or wandered instead to the Prater, where he befriended Arab fakirs—eaters of fire, swallowers of swords, tellers of tall tales, and his interest in exotic lands and tongues quickened. He knew Hungarian from his mother's talk; now he astonished people by the number of Arabic words he learned from the fakirs. He began to read books about foreign places, pretending he was an explorer in some faraway land.

The magic place for Joseph was China, and at the age of 13 he started to teach himself the language. He stuffed his pockets full of little cards with Chinese characters printed on them. The preoccupation irritated his father, who berated him for wasting his time with useless knowledge. But Joseph went on studying, reading by candlelight after the rest of the family had gone to sleep. He day-dreamed of Peking and Lhasa, but the dreams of his sleep scared him.

In school Joseph defended himself against the darts of his classmates with arrogance. They, with characteristic adolescent insensitivity, did not recognize a boy craving affection; they simply assumed they were being snubbed and took to calling him the "Count," after his father's employer. Joseph, mistaking their jibes for admiration, acted the part convincingly and became trapped in a myth of his own making. Meanwhile, the possession of an inquisitive mind was not, in itself, enough to endear him to his teachers. At the *Gymnasium* (middle school) instructors were not disposed to appreciate unapplied intelligence and they resented his truancies. They observed in him the potential for a star pupil, sighed after him when he produced work of only ordinary intelligence, and believed him to be either perverse or lazy.

Franz Rock had mapped out his son's future. Through Joseph he plotted to realize his own dream: the boy must be a priest. The homespun masses served as more than simple devotional exercises; they were training courses. Joseph had a phenomenal memory and, at an early age, knew all the rituals by heart. Moreover, he had a good ear and loved music; he charmed his audiences with sweet renditions of Gregorian chants. In many ways he enjoyed the sensation of power which his performances gave him within the family circle. But he had no intention of entering the priesthood. He wanted to get out of Vienna, to travel, to explore, and have great adventures. When he was small, his chatter amused his father, who dismissed it as childish fantasy. When the fantasy persisted, as witnessed by Joseph's stubborn study of Chinese, the old man began to complain. Lina, as usual, sided with her father; there were new causes for family squabbles, and Joseph learned to keep his dreams to himself.

As he began to comprehend the structure of the society into which he had been born, Joseph resented the accident that consigned him to the bottom of the social heap. In Vienna as in the rest of Europe, frogs did not change into princes. The son of a servant was destined to become a servant, marry the daughter of a servant or some appropriate equivalent, and propagate lots of little servants. But proximity to luxury—Count Potocki's quarters—influenced Joseph's tastes, and he took an early interest in money, not as a source of power but rather as an avenue to personal pleasure: fine furniture, tailor-made shirts, boxes at the opera, gourmet dinners. Under the spell of a sophisticated city and its aristocratic denizens, he developed urbane values and acquired that special worldliness that distinguishes a Viennese from an Austrian, a Parisian from a Frenchman, or a New Yorker from an American.

As Joseph's schooldays neared an end, the day of reckoning approached. He pleaded with Lina and argued with his father, but they held fast to their intentions to make him a priest. He begged them to let him join the Austrian Navy, but his father bellowed that it was the seminary or nothing.

His options thus clearly stated, Joseph chose to leave Vienna as soon as he received his *Gymnasium* diploma in 1902. He did not, finally, sneak off in the middle of the night, but paid a ritual farewell to Lina and his disgruntled father, who withheld his blessing. Joseph spent the next three years of his life wandering in Europe and North Africa, by sea or train, at his own leisurely pace. Lina secretly saved household funds so she could send him a few schillings now and then, but most of the time he took odd jobs as a guide for tourists or paid for his passage on ships by working as a seaman. He improvised brilliantly.

Two events stand out from those years of meandering. The first occurred in 1902 when Rock tried to pay the conventional call on Pope Leo XIII. Vatican protocol dictated a dark suit for the occasion and Rock, in his diminished circumstances, did not own one; he was not permitted to join the papal audience. Until that moment, though he had fought against becoming a priest, he had considered himself a Roman Catholic. But, wounded and insulted by regulations, he now turned away from the Church, and from that day on could never find a kind word for the formal trappings of any institutionalized religion.

The second incident was the death of his father, which happened while Rock was in Ostend in July, 1904. Lina sent a message and Joseph returned dutifully to Vienna for the funeral, his emotions badly scrambled. Franz Rock had been a poor man all his life and did not leave much for his heirs. Lina observed her brother's penury and gave him his father's gold watch and what money she could spare. Following a decorous period of mourning, Joseph resumed his travels and headed for England. There his health, which had always been frail, took a bad turn, and he learned he had incipient tuberculosis. When he began hemorrhaging, he panicked and returned home. Lina nursed him for a few weeks, but the damp, raw Austrian autumn only aggravated the disease, so he hurried south to Italy, then to Tunis, and wound up on the island of Malta, where he used what little money he had to rent a flat-roofed house with a garden on top. He convalesced in the Mediterranean sunshine.

The bleeding stopped; he felt improved, and his money ran out. He hired himself out as crew on a ship destined for Hamburg. This was not, as it turned out, a very intelligent decision. It was January, and as the ship moved north, icicles formed on the rigging and the decks became icy. Predictably, Rock fell ill again and had to stay in Hamburg, in and out of charity wards, until August. He left Hamburg, went to Utrecht and on to Antwerp where, once more, he landed in a hospital; when he was out of danger he was discharged again.

"One day," he recalled later in his dairy, "I bought a ticket to Aix-la-Chapelle but missed the train and sailed instead for New York on the same day at 9 a.m., I think it was the 9th of September . . . I remember how the spray of the sea on the quay in Antwerp was frozen and how hot it was in New York." He paid for his passage to the New World by working as a steward—a seasick steward at that—and docked in Manhattan without a penny. He left the ship wearing his uniform over his only suit and went directly to a waterfront pawnshop where he got fifty cents for the uniform.

Rock found a job as a dishwasher in New York City, and a new friend loaned him a bed. Anxious to learn English, he carefully avoided German-speaking people and caught on quickly. He spent his free Sundays in scenic graveyards, falling back on youthful habits and finding comfort among the tombstones of strangers. The arrival of cold weather triggered a new attack of consumption and he hastened to Bellevue Hospital for treatment; he improved again and, during the summer, was well enough to take menial jobs at resorts in the Adirondack mountains in up-state New York. But he was far from cured, and sieges of bad health depressed him. Warned by doctors against the climate of the northeastern United States, he decided to follow their advice.

He left New York by steamer, almost precisely one year from the day he had arrived, in search of some salubrious spot, vaguely destined for the Southwest. He sailed via Havana to Vera Cruz and dallied pleasantly in Mexico for a few months. The winter of 1906-07 found him in San Antonio, Texas, and during the summer he moved to Waco, where he found a part-time job and a kind family that rented him a cheap room. Hoping to polish his English, he enrolled at Baylor in two courses: English and Bible Introduction—religious study still being the most natural thing for him—and made plans to settle down for a time. But his lungs went bad again and, for no discernible reason, and contrary to the warning of a physician that the ocean air would kill him, he determined to go to Hawaii.

Rock bundled up his few possessions, set out for Los Angeles, and proceeded to San Francisco, still in ruins from the earthquake. He was never in much of a rush, so it was not until October, 1907, that he set sail for Honolulu aboard the *Manchuria* with one gold piece and some small change in his pockets. During the voyage a Chinese passenger organized a dice game and Rock, trying to double his small fortune, lost it in a few minutes. The luckless Chinese, who had staked much more than Rock, also lost everything, including "face"; he slit his throat before the ship docked. Rock felt an eerie tremor in the presence of the grisly death; more problematical, however, was the loss of his gold piece. He had gambled with his money for the first and last time.

He forgot his worries and bad luck as the *Manchuria* steamed into the harbor at Honolulu. The mountains of Oahu stretched toward him, reaching down like fingers into the sea; the scenery was dazzling, for Hilton and Holiday Inn had not yet invaded Waikiki. Rock debarked to the traditional cries of "aloha," which means hello, goodbye, good luck, and love. It was a word he would learn to use.

BEGINNINGS OF THE HAWAII HERBARIUM

Joseph Rock arrived 10,000 miles from his birthplace, a 5 ft. 8 in., fair complexioned, bespectacled young man, afflicted with a serious illness and unarmed with either money or academic degrees. For five years he had traveled through Europe and North America, alighting at random, at once in flight from his confused childhood and in pursuit of his health. Death was always breathing down his neck, and he never stopped anywhere long enough to form attachments, not even, it appears, to any representative of the opposite sex. His formidable memory stored the names of places and dates while disposing of people with the exceptions of a gentleman in Mexico City, whom he noted as once having been physician to the Emperor Maximilian, and a Mr. Lapp, who managed a paper factory in San Rafael. The habits of loneliness developed in Vienna were irreversible.

There were, of course, two ways of evaluating solitude, and even Rock could appreciate both views, though seldom simultaneously. While he had led what often had been an agonizingly lonesome life, he also unwittingly achieved perfect freedom. There was nothing to prevent him from packing his bags and leaving any place at any time; he had no apologies to make, no people who counted on him to provide either financial or emotional subsistance. Twenty-three tends to be callow, but years of fending for himself had molded Rock early. Behind a rather ordinary, gray-eyed exterior was a powerful will, a well-developed ego, and a little matter of nimble intellect to date evidenced only by his mastery of languages. Though friendless, he was friendly, impeccably appointed with Central European manners studied in the home of Count Potocki—and one adds, as a tribute to his good taste, that he never stooped to kissing the hands of American ladies. He conversed wittily, with twinkling eyes and infectious laughter as he regaled social gatherings with tall tales of his adventures; he listened with forbearance.

It is said that travel broadens, and no doubt Rock was broader for his experiences. For one thing, he had learned to express himself in a long list of languages. By the time he reached Hawaii he could cope more or less fluently in Hungarian, Italian, French, Latin, Greek, and Chinese; the Arabic he had picked up at the Prater in Vienna had improved on his visit in Tunis, and he could read Sanskrit. He spoke nearly unaccented English and, for the rest of his life, favored it over his native tongue, seldom lapsing into German. This linguistic smorgasbord might have impressed Ivy League academicians, but it did not make a dent in Honolulu, where people valued mercantile genius over cultural achievements. Yet it was here, in an alien cultural landscape on America's ultimate western frontier, that Rock chose to arrest his peregrinations for the time being.

Hawaii, as he found it, had been a United States territory for seven years; it had held out as a monarchy until 1893 when Queen Liliuokalani—an imposing lady who greatly feared witchcraft—was ousted. The Caucasians, or *haoles* to Hawaiians, though outnumbered by people of Polynesian and Oriental descent, already had a grip on the local economy. But neither the white man's industrious traits nor his Christianity could spoil the pagan beauty of the islands, which were at once a place for making fortunes and for luxuriating in sub-tropical comfort. Honolulu was the center of all commercial activity and, for the whites, of a limited social and cultural life. Ranchers and plantation owners from the islands converged in Oahu to conduct business, meet friends, and trade small talk. Their numbers being few and the society self-consciously introverted, the result was a provincial atmosphere in which no newcomer escaped notice.

Rock's immediate problem was money, and he solved it by taking a job as one of three full-time teachers at the Mills School, later the Mid-Pacific Institute, instructing Latin and natural history. He knew Latin from his Viennese schooling and religious exercises, and he improvised the natural history, investigating on his own to stay ahead of his students and grateful for the excuse to work out-of-doors. Besides being climatically and topographically unlike anything he had seen before, Hawaii had, and miraculously still has, distinctive forms of plant and animal life. Rock, therefore, had much to learn; but given incentives, there was no limit to what he could achieve, and he taught natural history very well. By summertime, however, his tuberculosis bothered him again, and he decided to find some kind of work that would keep him outside. He resigned from the Mills School in September, 1908, and a month later joined the Division of Forestry of the Territory of Hawaii as the first Botanical Collector, assigned to collecting seeds and herbarium specimens of rare Hawaiian trees and shrubs.

There are at least two extant versions of Rock's entrance into botany. One finds him convalescing on pre-tourist Waikiki beach, restlessly collecting algae to amuse himself. Alexander Hume Ford, a publisher and writer of independent means with an interest in things natural, and for a time director of the short-lived Pan-Pacific Institute of Botany and Acclimatization, espied him and volunteered to pay for the construction of Hawaii's first glass-bottom boat, thus launching Rock on his botanical career.[1] The second story—which does not necessarily cancel the first—pictures him storming into the office of the Division of Forestry demanding: "Do you have an herbarium? No? Well, you need one, and I'm going to collect it for you."[2] Ralph S. Hosmer, one of the foresters, took the responsibility, and later, when Rock became famous, the credit, for hiring him. Rock was lucky in his timing because the two Yale foresters then employed by the Division could not tell one Hawaiian tree from another. His idea that the organization should have an herbarium was not outrageous but, in the absence of credentials and training, his assertion that he was the man for the job demonstrated an uncommon measure of self-confidence. In any case, he had his outdoor job now and, as he had hoped, it put an end to his ailment. Furthermore, he lived up to his claim and began a first-rate herbarium.

There is no reason to demean Rock for his lack of academic background in botany; on the contrary, one marvels at his excellence in a science which he taught himself. Many botanists before him began as amateurs, though most of them, like John Torrey, Asa Gray, and George Engelmann, not to mention William Hillebrand, "Father" of Hawaiian botany, arrived to botany via medicine. Charles Sargent, the man who built Harvard's Arnold Arboretum practically single-handed, and for whom Rock later made an expedition, was one who dove into botany like Rock, with no specific scientific training. Rock attacked his work by spending as much time as he could with living plants in their natural habitats and, at night, by studying classical botanical literature as well as what was available on the Hawaiian flora. He conducted his field work with the ebullient enthusiasm and the persistence of a private eye tracking a missing person. Synopsis of a standard trek, as recorded in the Field Book at Honolulu's Bishop Museum:

"I rose at 4 o'clock. After a hardy [sic] breakfast composed of eggs, pancakes, pears, butter toast and coffee we were on the way to Mauna Kea, this time determined to find the beautiful *Ahinahina* of the natives, a very pretty composite plant called silversword by the foreigners. The air was just glorious, the sky was clear, with the exceptions to the east where clouds were hovering on the forest below . . . We went through the paddocks near Kaluana . . . towards an old extinct pit crater called Moano. I had visited the same place just about a year ago. There we rested our horses and ourselves." There followed some stumbling through old lava flows until "ah, here it is, the long sought *Ahinahina,* its beautiful silvery-grey contrasting [with] its background of old dead black lava."[3] The patterns of his later expeditions were already set: the early risings, the huge breakfasts, the reliance on native information, the stubborn pursuit. He loved the field and despised "armchair botanists," scientists who did their work exclusively from dried specimens and books.

Rock proved very resourceful in arranging land transportation. He could go to any island by boat but, once arrived, needed horses and guides in the country, and the Division of Forestry operated on too small a budget to provide him with money to hire them. He approached a rancher in the vicinity he planned to botanize and explained his mission; usually, the rancher, though ignorant of botany, was intrigued by the idea that someone thought the plants on his land worthy of study and volunteered horses and a couple of ranch hands for the effort. Very often, too, Rock was invited to stay at a ranch and so, in addition to learning the Hawaiian flora, he befriended many people on the islands. He soon earned a reputation as a desirable house-guest, particularly admired by the ladies who, consigned for long stretches to the boondocks, craved diverting company. Rock won them with witty stories and continental charm. It was during one of these visits, while staying with the Hind family on their ranch on the Big Island, that he was dubbed Pohaku, Hawaiian for rock or stone, by one of the native ranch hands. The little Hind girls picked up the new name, and it stuck with him.

Rock worked with the Division of Forestry for nearly three years, collecting, studying, and orienting himself in the islands. In 1909 he published his first paper, a description of a new species of *Scaevola*, and he prepared a forestry and botany exhibit for the Alaska-Yukon-Pacific Exposition, which won a gold medal. Though the Division of Forestry acknowledged the excellence of his work, its limited means forced it to reconsider priorities; it needed a forester more than a botanist, and Rock's inclinations were clearly more scholarly than practical. He had become a luxury the Division of Forestry could no longer afford. A friendly parting ensued in 1911. Rock easily found a new position at the College of Hawaii. The Division agreed to transfer the herbarium he had amassed to the College indefinitely, so that he could use it for teaching and for the continuation of his studies. He was made Consulting Botanist to the Division, an honorary position in which he did helpful work free of charge.

The College of Hawaii, parent to the present University, was established in 1907 as a land-grant college under the provisions of the 1862 Morrill Act. When Rock joined the faculty four years later, it numbered less than a dozen men, and the limited curriculum was weighted on the side of agriculture and engineering. Rock appeared in the 1911–1913 catalog as botanist. Long after his affiliation with the College had been terminated, he claimed to have taught Chinese as well as botany, but between 1911 and 1920, no catalog indicates that the College offered a course in the language. It is possible that he tutored students privately. Another curious point is that is 1914–1916, the College began the practice of identifying its faculty members by publishing the institutions which they had attended and listing their academic degrees. Rock's name is followed by University of Vienna without letters for a degree. His family in Vienna has no recollection that he ever went to the University; the University itself has no records of his attendance; and since he began his European wanderings at the age of 18, there would have been no time for him to have been a student. These discrepancies are worth mentioning only because Rock was extremely touchy about his academic credentials. By the time of his death everybody outside his family believed he had graduated from the University of Vienna and had been professor of Chinese at the College of Hawaii. Two young Hawaii botanists, Paul Weissich and Horace Clay, who were Rock's protegees years later, once raised his hackles by asking innocently where he had earned his doctorate. Rock turned red and would not speak with them for several days.[4] Actually Rock fabricated his university affiliations to save face among his academic colleagues and, in view of his demonstrated abilities, nobody ever thought to question them.

The study of systematic botany did not draw crowds, and Rock never had more than a few students at a time. He freed himself of normal classroom drudgery and did his teaching in the herbarium or the field, thus continuing his private investigations of Hawaiian vegetation. Edwin H. Bryan, who took Rock's course in 1919 and eventually became Curator of Collections and Manager of the Pacific Scientific Information Center at the Bernice P. Bishop Museum in Honolulu, recalled, not unkindly, that Rock made him a "slavey."[5] At the time Bryan was his only student, and Rock had him out collecting, copying Latin and German descriptions out of books at the Bishop Museum library, mounting dried plant specimens, and putting finishing touches on his papers.

Bryan worked side by side with Rock on the study of the genus *Plantago* in Hawaii, published under Rock's name in the *American Journal of Botany* in 1920. Bryan learned, and Rock was spared a lot of petty details in this symbiotic relationship. Working with Rock, however, required a quick wit and forebearance on the part of the student because the teacher was "as temperamental as a prima donna." Bryan learned how to read him: "If he were singing snatches of grand opera in French, German, or Italian when he arrived in the morning, the day would be a pleasant one." Silence, on the other hand, was ominous, signaling a difficult day filled with restless demands, moodiness, and barbed criticisms. The bad moods seldom lasted more than twenty-four hours in one stretch, but they were memorable. Bryan, an intelligent young man, was fortunate; a poor student stood no chance with Rock, for he had absolutely no patience with mediocrity or laziness. As sometimes happens with gifted people, he lacked understanding of, and any desire to understand, those who could not learn at once according to his instructions, and he refused to waste his time with them. But clever students like Bryan profitted in bondage. Rock lectured informally in either Latin or English, and Bryan transcribed the notes. During one year the two of them arranged and catalogued the entire herbarium in addition to the rest of their work.

Rock stayed at the College of Hawaii until 1920; in 1919 he was promoted to the rank of Professor of Systematic Botany, a title that no doubt did his heart good. However, it is clear from his bibliography for those years that professorial work took second place to botanical research. Between 1911 and 1921 he published more than forty-five studies in botany and forestry of which three were full-scale books: *The Indigenous Trees of the Hawaiian Islands* (1913), *The Ornamental Trees of Hawaii* (1917), and *A Monographic Study of the Hawaiian Species of the Tribe Lobelioideae, Family Campanulaceae* (1919), and more eloquent evidence of scholarly productivity is hard to find. These works resulted from concentrated physical and mental exertion over the nine year period.[6] Among Rock's botanical products, *Indigenous Trees* is the real classic, still indispensable to any student of Hawaiian botany. The volume earned him an international reputation among scholars.

In about a dozen years Rock achieved in botany what

would take most ordinary men a lifetime. The field work alone, given the rough and varied terrain, required stamina, and he seems to have known every square foot of the islands. People who joined him on field trips still marvel at his memory. Horace Clay and Paul Weissich, who accompanied him on several excursions in the 1950s after he had returned from China, note that he could remember the precise location of a tree he had not seen in thirty or forty years. If the tree happened not to grow there anymore, the fault lay with the tree, not with Rock. Beyond his systematic work he performed a service by recording the native Hawaiian names for plants, how the indigenous population used them, and their legendary properties. As a result of foreign travels, Rock introduced many new plants of economic importance, especially to forestry, and of ornamental value. Some of his finest introductions grace the campus of the University of Hawaii in Honolulu.[7]

ROCK BECOMES AN AMERICAN CITIZEN

Though Vienna had an emotional hold on him, Rock considered himself sufficiently at home in Honolulu to apply for American citizenship, and it was granted him in 1913. He used an Americanized form of his name. For the first time since leaving Austria he grew tentative roots and made friends. During his early years in Hawaii he boarded with Mr. and Mrs. Frederick Muir—he was an entomologist for the Hawaiian Sugar Planters' Association (HSPA)—near the College on Liloa Rise; Muir and Rock shared a love for opera and biology. Rock's closest friend was Harold Lloyd Lyon, a plant pathologist at the HSPA who arrived in Honolulu from Minnesota the same year he did. Lyon had a charming wife and, of course, the professional interests of the two men overlapped. They were friendly for years until the 1950s, when the relationship degenerated into a ridiculous quarrel over who had introduced which species of plants to the islands—a situation Lyon aggravated and Rock took too seriously. There were the people with whom Rock stayed when he was collecting outside Oahu, like the Hinds or Herbert Shipman, a wealthy rancher with horticultural proclivities, who encountered Rock on Hawaii, one fine day in 1908, looking for a black wasp.

Yet Rock was still a loner. An intensely private man who gave very little of himself, his friendships lacked intimacy. Prone to moodiness, there were days when he felt gregarious and sought society; other days he fled it in a panic. He maintained his friendships on a relentlessly formal level, even by European standards. After more than three decades he still began letters to Harold Lyon with "My dear Dr. Lyon," and signed them "Cordially and faithfully yours, J. F. Rock;" Lyon was less than five years his senior. With his contemporaries, his casual form of address was to use their last name; sometimes he signed J. F. Rock, or Rock, or in a few cases Pohaku, usually supplying the J. F. Rock in parentheses after his Hawaiian nickname, lest there be a misunderstanding. In his native German he could have accomplished the same aloofness by using the polite *Sie* instead of the familiar *Du*. Consciously or not, his insistence on formalities defended him against—or denied him, depending on one's point of view—the ultimate exchange which bind friends in love and also make them vulnerable. He kept his freedom intact, and used it. But freedom had a price, and Rock paid it: "I was never happy [in Hawaii]," he recalled later, "in spite of all my friends; in fact, I was dreadfully lonely."

Rock stayed put in Hawaii for more than five years, botanizing, writing, teaching, and living frugally. Restless by nature, he worked out his excess energy in his exploration of the islands, where nature supplied enough variety to divert him. Then, in July 1913, he went on a two week field trip with Judge Henry E. Cooper and C. M. Cooke to the miniscule atoll of Palmyra, south-southeast of Hawaii, an expedition sponsored by Cooper, who had purchased the islets and wanted to find out what he had bought. "I frankly asked the Honorable Judge . . . what made him buy these islands long before he had ever seen them. He answered that he had always wanted to possess a yacht and a place to go."[8]

The yacht *Luka* was a 70–ton, 70–foot craft with a glorious past and a doubtful future, captained by a "German sea hawk [who] made it plain to me that there was not the slightest distinction between him and the captain of the *Mauretania*." The sea voyage, punctuated with the kind of mishaps that seemed humorous only in retrospect, brought them to the islets, wretched specks of land overrun with hermit crabs and impudent booby gannets. The three men spent sixteen days fighting the elements and collecting plants and animals. Years later, in a witty piece for the *Atlantic Monthly*, Rock wrote: "What finally happened to Palmyra I have not learned . . . But once it nearly became the property of a group of Chicago ladies, man-haters who wanted to retire to this lonely spot in the mid-Pacific, but had they ever ventured there, I would vouch that they would have turned woman-haters in the end."

Rock arranged for a leave of absence from the College in September and, using what he had saved, began what his earliest existing diary refers to with charming childishness as "my trip around the world." On the 14th the day of his departure, he took a final swim at Waikiki, rushed up to his lodgings at the Pleasanton Hotel to pack his bags, and went down to dockside where half a dozen friends, including Hosmer and Lyons, were on hand to see him off in Hawaiian style. Standing on the deck as the *Thomas* steamed out of the harbor, his neck wreathed in sweet-smelling leis, Rock felt his emotions play tricks on him and said goodby sadly "to the mountains which are so dear." The delights of travel, however, quickly preoccuppied him.

The *Thomas* stopped first at Guam. Rock had a day on the island, made methodical notations on vegetation, and boarded again. At the next stop, the Philippines, Rock received a two-week low-key introduction to the Orient. He checked in at the Army and Navy Club in Manila and made a quick survey of his surroundings. In native sections of the capital and in outlying villages he saw wretched deprivation and was torn between pity and disgust. He found a new acquaintance in Elmer Drew Merrill, botanist at the Bureau of Science, and spent several days collecting plants and seeds in the forests, hoping to find species suitable for the reforestation of Mauna Kea and Haleakala, a project of Hawaii's Division of Forestry.

Rock left the Philippines on October 19. Two days later he landed in Hong Kong, noting this arrival in his dairy with the Chinese characters for the city, which he had learned during his childhood. He went from ship to hotel by rickshaw, commenting on the "funny feeling to be carried by a human being," which would one day seem quite ordinary to him. The city pleased him. He shopped, spent too much money, botanized casually on the hills of Kowloon, and after three days decided to go to Canton. A boat took him to Shameen Island, the foreign concession, where he was met by a pleasant, English-speaking guide. Borne on rickshaws, they passed the Chinese police guarding the half-English, half-French compound and entered the Republic of China for a seven and a half hour tour of the city. "It is the most interesting place I have ever seen and hope to see," claimed the excited Rock after the long day. A rickshaw tour of Canton was not precisely what he had dreamed about, but it was tantalizingly close.

"Our coolies cry out constantly as we are carried through the narrow streets, 'make room for the foreign devil,' continually repeating the above phrase . . . a few boys are angry at me, kick my chair . . . and then stick their tongues out at me as well as curse me, but in Chinese." All was not well between the Chinese and the Europeans, but Rock would not be rattled. Having arrived on the site of a childhood dream, he marveled at its texture and activity, likening it to a beehive and a bazaar. He took it all in: crooked narrow streets; noxious odors of exposed sewers; tiny, filthy shops where people produced goods of improbable beauty in silk, ivory, jade, or embroidery; a waterclock which had kept time since 1325. The guide ushered him to five temples, including one dilapidated structure that honored Marco Polo among its five hundred idols and another—called the Temple of Horror—which seemed to Rock like the "congregating place of the scum of Canton," and from which he retreated hastily after snapping a few photographs. When the tour was over, Rock returned to his ship for the night voyage back to Hong Kong. That night he wrote in his diary the longest, most enthusiastic entry of the entire trip. He had not been disillusioned.

Everything after Canton seemed pedestrian, and Rock's diary became sketchy, outlining the course of his travels with a simple modifier here and there. Singapore was "pretty," Darjeeling "beautiful," India and Ceylon "interesting," but China had captured him. He sailed from Ceylon to Europe, his head stuffed with Oriental images.

He went especially to look through Hillebrand's herbarium at the Botanisches Museum at Berlin-Dahlem because Hillebrand thoughtlessly had not left any specimens in Hawaii. Rock secured a thousand sheets of fragments and duplicates of Hillebrand's types to send back to Honolulu. This eventually proved a fortuitous acquisition since the Berlin herbarium was destroyed by bombs during World War II. Back in the United States, Rock visited Harvard's herbarium to pick up a few more specimens for Hawaii.

Rock returned to Honolulu and resumed his teaching and scholarly writing. But whenever he had adequate funds, he traveled again, always using plant introduction to justify his wanderlust. He went to the Philippines and Singapore again, stopping at Java, on a trip funded in 1916 by the HSPA; in 1919 he revisited Siam, Malaya, and Java. He did field work in southern California in 1917. He produced no diaries for these years, a misfortune because, aside from the desirability of tracing his voyages, it would be interesting to know his feelings about the war that ravaged Europe, but in which he had no part.

Bryan, Rock's student, remembered that "the feeling that Mr. Rock was not happy with his position at the College of Hawaii became very apparent. On November 26, 1919, my diary records our completion of the paper on *Plantago* and 'he says he is going to leave.' "[9] Rock's malaise coincided with the transition of the institution from college to university level. The change involved an agreement between the University and the Bishop Museum designating the latter as the depository for systematic collections. Effectively, Rock's herbarium, by then some 28,000 specimens, was to be moved to the Bishop Museum. While the herbarium was not his private property, it had become a very personal thing to him and he felt affronted that a bargain had been struck over his objections. Needless to argue that the college was pressed for space or that the new arrangement would benefit the Museum; Rock inferred insults to his integrity and the dignity of his work. The legislation authorizing the move was passed in 1920, and though the actual transferral of materials did not take place until 1922, Rock left the College and Hawaii in a huff on May 25, 1920, and headed for the mainland to peddle his talents.

Specialists in Hawaiian botany were not exactly in demand at the time. Mindful of the Gray Herbarium's history of interest in Hawaii and the Far East, Rock went to Harvard looking for a job, but the herbarium led a hand-to-mouth existence and could not afford to add to its small staff. From there he presumably tried the New York Botanical Garden and met with similar results. It was in Washington, D.C., that he finally hit upon a good thing. The Office of Foreign Seed and Plant Introduction of the U. S. Department of Agriculture was interested in importing seeds of *Hydnocarpus kurzii*, the chaulmoogra tree, a native plant of southeast Asia, which produces an oil useful in the treatment of leprosy. To achieve this, they needed someone who could recognize a chaulmoogra tree when he saw one, who knew how to pack and ship seeds—a task more exacting than it sounds —and, preferably, who had some familiarity with the flora of the area in which the tree might be found. Rock was made to order for the job, and it for him. He snatched it. By the fall of 1920 he was on his way back to the Orient as an Agricultural Explorer.

SEARCH FOR THE CHAULMOOGRA TREE

As a representative of the United States Government, Rock arrived in Bangkok armed with credentials and letters of introduction. He had already been there in 1919 and loved the city for its fanciful wats, which fingered the sky. His mission foremost in his mind, however, he postponed tourist pleasures and after checking into his hotel, called at the American Legation to solicit the good will of the representative.

"When sending in my card I was ushered into a large room to a central desk at which sat a corpulent man, with a round red face and a yellow mustache, the ends of which were waxed to two sharp points. He greeted me in a very democratic manner, took the letter of introduction and at the same time motioned me to take a seat opposite him at the long desk. He smoked cigarettes and, opening a silver case, said 'Have a cigareet,' which expression gave me a jolt, for I was not prepared for such a breach of English from the mouth of a Minister Extraordinary and Plenipotentiary, but soon I was to listen to more such democratic expressions from the Minister.

"Some correspondence had passed between the Legation and Washington in regard to my visit to which I called the Minister's attention, whereupon he rang the bell for the Siamese interpreter, a certain Mr. Leng Hui, who was part Chinese and part Siamese. As he entered into the presence of the Minister, he stopped and bowed and said 'Mr. Minister,' and waited for the command of the man at the desk . . . With a chuckle which forced him to withdraw his short neck still deeper into his collar, the Minister folded his arms over a broad bosom and, looking at me, said, 'Do you feel any better for that bow?' 'Ha, ha, ha,' he roared, 'Mr. Minister! ha, ha, ha. In my home town they call me *George*. Ha, ha, ha.' Motioning to the embarrassed secretary to draw nearer, he took him by the hand and said, 'Say, if you ever get out of this country, don't bow to anybody.' "

George Hunt, an affable Arizona grocery man, politician, and one of President Wilson's less inspired appointments, had an interest in agriculture and befriended Rock, inviting him to stay at the Legation. Rock accepted, thus surrendering himself to a month of alternating hilarity and embarrassment. Completing the family were Mrs. Hunt, whom Rock remembered as a woman who had lost four fingers in a haying machine and equally bereft of diplomatic graces, and an unappealing 14-year-old daughter.

When Rock wanted to obtain cuttings of an especially succulent grapefruit for American citrus growers, George insisted on accompanying him. "The [Siamese] Government very thoughtfully sent a member of the Siamese department of agriculture, who turned out to be a graduate of Cornell University, to be our guide and interpreter. We had hardly seated ourselves and the boatman released the launch from its mooring when Mr. H., like a bolt from the blue, addressed himself to the Siamese official saying, 'Your king ain't no kind of a king anyway. He is only interested in them poetries. He cares a damn about agriculture.' " From table manners to modes of dress to political pronouncements, the well-meaning minister was a disaster, much to the chagrin of the entire foreign community, particularly the Americans. After a few weeks Rock announced he was leaving Bangkok for Chiang Mai in the north. George fancied the trip and made advance arrangements for a properly ceremonious reception. Forewarned, the American community—doctors, missionaries, and a representative of the Rockefeller Foundation engaged in a campaign against hookworm—prepared a luncheon in the minister's honor. His behavior so horrified them that one of the hosts whispered to Rock, "We wish to God he hadn't come." George, however, remained childishly insensitive to his own *gaffes* and said after the festivities: "A Minister Extraordinary and Plenipotentiary doesn't come up here every day. Gives them fellows a sort of standing now."

Rock, who never quite recovered from George, classified him as a "square plug in a round hole." Shocked by what he interpreted as backward ways, this protoypical Ugly American was full of good intentions and democratic platitudes; he was also totally ignorant of Oriental forms, customs, and sensibilities. Such innocence, both at home and abroad, probably did more to undermine the United States position in the Far East than any gunboat or economic exploitation. Though this was Rock's first and most stunning encounter with inept diplomacy, he was already sufficiently sophisticated in his comprehension of things Oriental to realize the damage that a George could do. This George, fortunately, lasted only a year, probably to his own, and the Siamese government's relief. His parting shot, incidentally, was a letter of recommendation for one of the household servants which read: "This is to certify that Ah Lum has been at this legation ever since I have been here as cook."[10]

"Diverting though his month with the American Minister had been, Rock had found time to reacquaint himself with Bangkok's wats and to glean information from local people concerning the likely locations of *Hydnocarpus kurzii*, for the chaulmoogra oil had long been used as a native remedy for leprosy. He left George in Chiang Mai on December 2 and began his search, descending the Mae Nam Ping on a commodious house-boat with a Lao crew, a cook, and a servant. The ten-day river journey bored him except for occasional sorties to the river bank where he could collect plants. Once, exasperated by the sluggish progress, he dove into the river and started swimming ahead of the boat; the Lao crew stood on deck, waving and shouting something he could not understand. Later he learned that the river was infested with crocodiles. Neophyte explorers often stay alive by luck; Rock would become progressively careful with experience. However, he began asserting himself with the natives

47

and, after one fragrant occasion, ordered the Laos to refrain from eating anywhere near him the rotten fish which they so loved.

At Raheng (Tak) Rock left the river and began a trek, with a small army of coolies, overland to Moulmein in Burma. In the Kawkereik hills along the Burma-Siam border he spotted a lone chaulmoogra tree for the first time but it was not bearing fruit. He reached Moulmein by Christmas Eve and passed the day pleasantly among missionaries. He found chaulmoogra seeds between Moulmein and Rangoon in a fly-ridden marketplace in a village. "Almost every grain is moving, and this in the midst of squatting, betel-nut chewing, and expectorating women." The charm of such novelties was wearing thin, but no one seemed to know the source of the seeds. Next he saw *Hydnocarpus castanea,* a closely related and, for his purposes, useless species. Finally, at Kyaikto, he discovered the true chaulmoogra, fruiting heavily, and ordered his coolies to collect the seeds.[11]

Rock continued the chaulmoogra search on into Bengal because one batch of fruits did not suffice; seeds gathered in different locations were bound to have different responses to growing conditions in different parts of the United States. A thorough agricultural explorer sought as many distinct seed lots as possible, and Rock was meticulous. In addition, he sent other valuable plants as he discovered them. Meanwhile, the USDA advised him of their interest in ornamental species, particularly a blight-resistant chestnut, which might be found in western China, and Rock indicated his willingness to go after them provided he could make a trip to Europe for recuperation. Flushed with success and with the money he had earned from the USDA, as well as the proceeds from an article he had written for the *National Geographic,* he set off for Vienna, the Prodigal Son returned.

Home again after almost twenty years, he played the irresistable role of poor-boy-who-made-good-in-America-visiting-his-less-fortunate-relatives and established himself in a plush hotel. Lina, divorced from her husband shortly after the birth of a third child, improvised a threadbare existence for herself and her sons. Rock gave generously, but his gifts smacked of American foreign aid programs in more than one sense. In the conviction that country life was better for the boys, he insisted on buying a little house for her in the country where she could set up a small shop. Lina, a thoroughly urban creature who loathed the idea of country living, had no choice but to take what was offered. Meanwhile benevolent Uncle Rock showered the boys with presents, drove them around Vienna in expensive taxis, and told them hoary, funny tales of his childhood and China. A cloud momentarily covered his visit when their taxi hit a woman one day, nearly—but not quite—killing her; Rock personally lifted the victim from the pavement and escorted her to a hospital. Otherwise, he enjoyed himself royally, particularly the night when he purchased a ticket for the Emperor's box at the opera. For a few hours money came gorgeously close to buying happiness.

Interlude over, late November or early December of 1921 found Rock back in Bangkok calling on the notorious George only to learn of his recent recall. Holding the fort while awaiting the new Minister was a *chargé d'affaires*, "a perfect drawing-room Gentleman, the very opposite of Mr. H., who was, of course, the subject of much of our conversation." Then, on December 30, Rock took up his diary again, in the back of the same book he had kept in 1913, to record his entrance into West China.

WESTERN SCIENTISTS DISCOVER CHINESE VEGETATION

Now that Rock was committed to botany, China offered him rewards beyond the realization of a boyhood fantasy. During the last forty or fifty years the western provinces, especially Hupeh, Szechwan, and Yunnan had been discovered by the West to be one of the richest botanical fields in the world. For centuries Western visitors had remarked upon the vegetative wonders of China. Marco Polo noted exotic, tempting fruits in the market places and exquisite flowers in Chinese gardens, and subsequent visitors—missionaries, traders, or government officials restricted to coastal areas—brought back further tales and sometimes a few seeds, scraps of plants, or even herbarium specimens. Quite a number of showy "florist flowers" arrived in Europe via the spice ships. Missionaries, less concerned with ornamental qualities, provided scientific observations on the native uses of plants, particularly for medicinal purposes. But, on the whole, China shut out the West until the 19th century when, defeated by Western military ingenuity and technology in the Opium Wars of 1842 and 1860, the Chinese grudingly made the concessions that made foreign penetration of the interior possible, if hazardous.

The West set a precedent of botanical investigation in China, observed initially by amateur and later by trained botanists, for commercial rather than scientific reasons. Exploration was prompted by curiosity and greed combined in uncertain proportions. The first large-scale botanical expeditions originated in England with the backing of such organizations as the Horticultural Society of London and the East India Company, which sent Robert Fortune on successive voyages in 1843, 1848, and 1852. The Horticultural Society ordered him, with refreshing naivete, to bring back ornamental plants as well as such potential money-makers as "the plants that yield tea of different qualities," "The plant that furnishes rice paper," "The canes of commerce," and "The varieties of Bamboo."[12] The East India Company charged him with collecting tea seeds in China for cultivation in Sikkim and Assam, a plan designed to circumvent Chinese taxes on the tea so beloved by Englishmen. Fortune ventured "inland" —at least inland compared to what Westerners had done in the past; to the tea districts in Anhwei and to Fukien. His success, with its financial rewards to his sponsors, did much to encourage his countrymen. The British proceeded to assault China's vegetation with pragmatic efficiency.

Yunnan, which would be Rock's home proince, became known to the West in the late 1860s when the British were looking for an overland trade route from India through Burma to China, via Bhamo on the Irrawady. In 1868 Sir Percy Sladen, an envoy of the Indian Government, got as far as Tengyueh in Yunnan before being turned back. That same year another official, Thomas Thornville Cooper, took the long way 'round, entering China from Tibet (Lhasa to Tatsienlu-Kangting); but in Weisi, Yunnan, he was arrested and barely got out to Burma alive. Progress on the India-China route was negligible. In 1890 Archibald Little described the road between Tali and Bhamo as "a footpath only passable by mules and pack-coolies, and on which mounted men are often compelled to dismount and lead their animals a great part of the way."[13]

While the British poked around the perimeters of China with mercantile motives, the French, vying for political and economic favors, stumbled upon scientific information in the deep interior. Gabriel Eugene Simon, an agricultural botanist in government service, traveled extensively through Anhwei, Hupeh, Honan, and Szechwan, collecting dried and living material along the way. Père Jean Pierre Armand David, a Lazarist missionary and innate naturalist, made important collections of birds and dried plants in Szechwan, visiting many localities that Rock would re-explore. Père Jean Marie Delavay, inspired by David whom he met in Paris, gathered over 200,000 plant specimens in northwest Yunnan in the 1880s and thoughtfully sent seeds of the choicest species to Paris where they were not always very well grown. Meanwhile a steamer route into Yunnan via the Red River from Hanoi to Laokai opened up in 1889 and gave the French a commercial edge in southwest China, irking the British.

Russia, by virtue of a common border with, and political designs on, China produced the third significant contigent of botanical explorers during the 19th century. The Russian naturalists—sometimes, coincidentally, military officers as well—penetrated the northern boundaries and collected in the northwestern provinces. Captain Nicolai Mikhailovich Przewalski made four excursions into China, to the Tibetan borderlands, the sources of the Yellow River, and the northwestern steppes, dying while planning a fifth expedition to Lhasa, the goal of all his journeys, which he never reached. Grigori Nikolaevich Potanin went as far south as Szechwan. Rock would also cross the tracks of these men. Another Russian, Emil Bretschneider, explored the hills near Peking where he was stationed as physician to the Russian Legation from 1866 to 1883. He is more celebrated, however, for his historical rendeing of Westren botanical exploration in China.[14]

Once the Chinese had relented to the hateful foreigners' demands for access to the interior, Westerners trickled inland, but comparatively few of them reached the western provinces. Among those who did—many, as noted, with botanical interests —were missionaries, traders, government officials, physicians to care for foreign communities where they existed, and members of the Chinese Customs Service, *i.e.*, foreigners in the employ of the Chinese government, assigned to collect customs tariffs. Augustus Margery, a British consular official, completed the first journey from the Yangtze to the Irrawady in 1867 but was murdered forty-five miles east of Bhamo. E. Colborne Baber, who eventually became the British Consul in Chungking, explored in Szechwan and Yunnan. His successor, Alexander Hosie, also under government orders, spent three years, between 1882 and 1885, traveling about to see what was to be seen of interest to Her Majesty. All three men contributed a smattering of botanical information. Antwerp Pratt, zoologist and entomologist, may have been unique in the singularity of his motives; he left England in 1884 with the sole intention of investigating the wildlife in China's far west. Augustine Henry, a Customs Service officer, languished first in Ichang on the Yangtze and later in Szemao in Yunnan; he started collecting dried plants to break the monotony of his existence. Henry's specimens, which he sent to the Royal Botanic Gardens at Kew in the 1880s and '90s, reawakened the interest in western China as the site of horticultural delicacies.

As a direct result of Henry's work, the prestigious Royal Exotic Nursery hired Ernest Wilson at the beginning of the new century; he was to secure seeds of *Davidia involucrata,* the charming dove tree originally discovered by Père David. Wilson finished by making four expeditions to Western China; two for the Royal Exotic Nursery and two for the Arnold Arboretum of Harvard University, introducing hundreds of new plants into European and American gardens and demonstrating that the flora of the area was incredibly rich. Wilson's success on his first effort marked the beginning of serious, large-scale botanical collecting in western China. The rage for rare Chinese plants started in England, where horticulture was a fashionable pastime, and spread to the United States. Close on Wilson's heels came George Forrest, a businesslike collector financed by a British gardening consortium; Reginald Farrar, and Frank Kingdon Ward. Both Forrest and Kingdon Ward covered much of the same country that Rock later collected; in fact, as Rock approached Yunnan in 1921–22, Forrest was afield near Likiang. All three men would meet and become wary of one another.

III

POLITICS AND LIFE IN YUNNAN, SOUTH OF THE CLOUDS

Until the equivocal appearance of the communist regime, Americans thought of China as a huge yellow expanse on a public school map, where the British Empire was colored red. Dalliance in Chinese history produced a few place names, a couple of wars (Boxer and Opium, easy to remember), a temporary hero or two: American educated Sun Yat-sen; Christian Generalissimo Chiang Kai-shek, who rose and then fell in public esteem; Chairman Mao, who has his selected partisans, and a mental picture of quaint yellow fellows in pointed hats knee-deep in sculptured rice paddies. After 1949 and during the Korean War, when the rice-growers seemed to have beaten their plowshares into swords, what used to be picturesque became menacing to the mind's eye. For reasons best known to the architects of American foreign policy, contacts with the new foe were confined to clandestine, low-level conferences in the Polish countryside, and the American public resigned itself to knowing less, rather than more, about China. Such formidable ignorance has generated convenient myths, the most popular one being that the Chinese are inscrutable.

There exists also a prevailing notion of China as a uniform geographical and cultural entity, something which is no more true of China than it is of the United States. Marco Polo found so much contrast between the Chinese north and south that he called them by separate names, Cathay and Manji, respectively. Dividing the country roughly along the 34th parallel, between the Yangtze Kiang and the Hwang Ho (Yellow River), that is, between the rice-growing and non-rice growing regions, a strong case can be argued for distinctions in climate, agriculture, the structure of cities, physical characteristics of the population, language, and even political attitudes. George B. Cressey observed in *China's Geographic Foundations,* that the "Chinese of Shantung and Kwangtung have little more in common than the French and the Italians and might have equal difficulty in understanding each other." One could also compile an impressive list of differences between the east and west of China. Europe and America's failure to recognize China's varied texture was the offspring of China's xenophobic reluctance to be scrutinized by outsiders and the West's egoistic idea that Chinese history began—in terms of anything that mattered—only with its arrival and continued only under its influence. This initial mismatch of attitudes, fatal to mutual understanding, successfully resisted the best efforts to set the record straight. The myths still prevail despite the initiatives of ping pong players, Henry Kissinger and Richard Nixon.

The western provinces of China have suffered most from intellectual neglect. When the Japanese forced Chiang Kai-shek to withdraw his government to Szechwan, Americans rushed to their atlases to find out where Chungking was. According to the redoubtable Edgar Snow, who traveled with Rock in Yunnan in 1931, the far west came as a surprise even to many Chinese. "In those travels," he recorded in *The Battle for Asia,* "I did not meet a single 'outland' Chinese, and afterward, back on the coast, I never encountered a Chinese intellectual who had once visited that magnificent part of his country. . . . Here, to Chinese school children, is the home of the most exciting myths in that wonderful classic, the *Shan Hai Ching,* which tells of headless men, people with perforated chests, and the desert of moving sands and the wilderness beyond the Northwestern sea." This was Rock's country, from Yunnan to Tsinghai, a land of mountains that rival Himalayas, foaming rivers, jungle in the south and, beyond Lake Kokonor—the Northwestern Sea of

Chinese folklore—windswept steppes, all unknown in America and Europe save for the writings of a few hardy adventurers.

"It must not be imagined that China is the aboriginal home of the Chinese," wrote Rock, "no more so than America is the aboriginal home of the present day Americans. It is true that while Americans count their occupation of that country, or some parts of it in hundreds of years, the Chinese count theirs in many thousands. Both are immigrants in their respective countries; the difference is only in time. Their immigrant status is recorded in their history and by the presence of still numerous aborigines, especially in the west of China. The Chinese have been less aggressive than the white man, and less thorough in killing off the aborigines."

China climbs from east to west, culminating in the formidable mountains of the Tibetan borderland. The aborigines of Rock's account in ancient times had lived on the plains and been driven to the mountains or jungles or, like the Nakhi, had been contained in the mountains. Owen Lattimore, one of the few Western scholars to take a serious historical look at China's "frontier" population, designated them as "refugee and remnant peoples crowded against and into Tibet by growth of the Chinese in the richer country."[1] Rock observed that even where there were plains among the peaks in the western provinces they were occupied by Chinese except where the land was too high for agriculture as among the grassy plateaus of the northwest. Tribes accounted for roughly 10% of the lowland population in Yunnan and 67% of the mountain inhabitants; in all, according to Cressey, there were about 200 tribal divisions in the province. Tibetans and Mongols dominated the northwest plains of Sinkiang, Tsinghai, and Kansu which, as Rock had noted, did not interest the Chinese. The western provinces were also the homes of large numbers of Mohammedans, descendants of a ninth century immigration from Turkestan and Persia, whose relationship with the Chinese was bitter. Chinese living in the west referred proudly to their origins so as not to be mistaken for tribesmen. Paul Meyer, former American Consul in Yunnanfu, recalled a man who said he was from Anhwei province. "How long have you been here?" asked the American. "Ten generations," replied the Chinese.

Like a true blue American (which he was not), Rock sided with the underdog tribesmen: "The Chinese . . . left the aborigines alone after they had once been chased from the rich plains to rule themselves under their own chiefs, except that the latter were nominally under some Chinese magistrate often many days' journey distant. With the advent of the Republic and Chaos the situation changed; they were treated as vermin, exploited and squeezed by the officials, while formerly they were more or less ignored and dismissed wtih derogatory terms. They are mentioned in Chinese history and literature, but usually after women, who had little standing in China. Most of them were only briefly dealt with in a line or two which usually ended in the phrase that they eat hair and drink blood. When they are figured they show their bodies covered with hair to denote them as savages, and since the white man is also hirsute he is mentioned in the great geographies of China after the savages, at the very end under tribute-bearing countries as *Hung-mao Fan-jen,* or red-haired barbarians." In his indignation, Rock neglected to mention that the hapless Chinese peasant, whose vanity the historians protected, was as surely a victim of bad government as the aborigine, just as the poor white in the southern United States had suffered as much under the slave system as the black.

Though the western provinces were indeed in a state of turmoil during Rock's residence there, chaos predated the Republic by an undeterminable number of years. In 1869 Abbé David recorded "it is said to be dangerous to travel [in Szechwan near Chengtu] . . . I am told about thefts and murders committed by bands of brigands."[2] He soon verified these rumors. Ernest Wilson, trying to get to Szemao in 1899, was stranded in Laokai (now North Vietnam) while Chinese soldiers chased a gang of outlaws who had burned the Customs House and French Consulate and assaulted foreign residents in Mengtsze. The foreigners supposed the attackers to be Boxers; they were, in fact, just opportunistic bandits. Under ordinary conditions the Chinese soldiers would have given up their pursuit quickly, but the red-haired barbarians put pressure on the district magistrate and, after several weeks, the wrong-doers were duly apprehended, beheaded, and had their heads mounted on poles in Mengtsze for all to see.[3] The great Mohammedan rebellion in Yunnan, Kansu, and Shensi began in 1855 and continued for eighteen bloody years, claiming perhaps ten million lives, according to Cressey. Fighting among tribes, or between tribesmen and Chinese, had been going on for centuries. On the whole, it is extremely doubtful that the western provinces ever enjoyed long periods without bloodshed, the distinctions between order and chaos being simply a matter of scale.[4]

From another point of view, however, Rock's equation of Republic with Chaos was not unjustified. He arrived in China at a moment when governmental authority had reached a nadir, and infighting among factions in the east magnified the disorders of the west. The political muddle in Yunnan alone was enough to turn the most ardent supporter of Chinese democracy into a cynic. This province had the longest history of rebellion against Chinese domination and was the last to be absorbed into the empire. From ancient times its record was marked by tribal feuds and revolts, and the Chinese seldom exercised more than nominal control. *

The Ch'ing dynasty had been hard pressed to keep Yunnan in line. Like the other western provinces, Yunnan's ethnic complexity and remoteness stood in the way of Peking's authority. Yet the Manchus had successfully pacified the Moslem uprising —the price in human lives, appalling by any standards, did not make the Ch'ings flinch—and had ignored lesser outbreaks of lawlessness except, as in the Mengtsze affair, when some outraged foreigner insisted upon his extraterritorial rights. As long as the governor delivered on time the taxes he squeezed from the province, the dynasty was satisfied. Under the Republic, however, Yunnan declared itself independent and passed from one warlord to the next.

Yuan Shih-k'ai, the Republic's first president, was so preoccupied with dreams of monarchical grandeur that he lost his grip on Yunnan and several other provinces and provoked his reluctant supporter, Sun Yat-sen, to rebellion. Yuan's obsession with power eventually carried him to his own destruction, which

---

* For further synopsis see *La Province du Yunnan* by G. Cordier (Le-Van-Tan, Hanoi, 1928). Parts of Yunnan are still designated Minority Autonomous Districts, including the Likiang region. See map in Edgar Snow's *Red China Today* (1971).

would not have been so tragic had he not, during the four years of his misguidance, brought the Republic down with him. Yunnan's break with Peking came on Christmas Day, 1915, under the leadership of Ts'ai Ao, a former military governor. Yuan, who had never trusted Ts'ai, called him to Peking in 1913 and held him there as a captive for over a year. Ts'ai escaped and swore to fight Yuan and save the Republic. When the president refused to yield to his demands, Ts'ai proclaimed Yunnan's independence, backing his words with an army of 10,000, which Yuan's generals refused to fight. Within five months of Ts'ai's declaration, warlords in Kweichow, Kwangsi, Kwangtung, Chekiang, Shensi, Szechwan, and Hunan had followed suit. Yuan's henchmen, quick to desert the sinking ship, left him alone to contemplate his shattered country and ambitions and, soon after, to die. In his wake he left an embattled nation.

Rock entered Yunnan in 1922 just in time to see for himself how China arrived at political decisions. In Szemao en route to Likiang he learned that Yunnanfu had been seized by a mob of 20,000 bandits and Ku Pi-ch'en, "the most decent and honest Governor the province ever had," had been shot. T'ang Chih-yao, Ku's successor, was a product of mission schools, Japanese military academies, and China's revolution; he had been one of Sun Yat-sen's earliest and most powerful supporters. Now, however, he had strictly personal objectives. He formed an army of outlaws and malcontents which carried him to power in Yunnan. Once there, T'ang found the army cumbersome and threatening. Dividing to conquer, he kept 10,000 of the best men for his army and dispersed the remaining 10,000 to resume their brigand life, promising to leave them in peace if they would not rob in Yunnanfu. Accordingly, Rock testified, bandits infested the countryside within a radius of three days from the provincial capital.

Owing to his long relationship with Sun Yat-sen. T'ang was a powerful national figure. In 1924 Sun appointed him Deputy Generalissimo, but T'ang declined the honor. When Sun died in March, 1925, however, the Yunnan *tuchun* marched toward Canton with the idea of assuming his post and power. Rival warlords defeated his armies in Kwangsi and the armies of his supporters in Kwangtung. His grandiose ambitions subdued and his army battered, T'ang withdrew to wheel and deal in his own province.

"After his self-appointment as governor warlord he had bled the treasury white and he needed money," wrote Rock. "The simplest way to get it was to put his [seal] on pieces of paper with the denominations 1, 5, 10, 100 dollars written in Chinese ink. The gold reserve behind it was bullets, for when the new money made its appearance and people were skeptical and refused to accept it, a proclamation was posted on the *Chin-ma* gate notifying the people that anyone refusing his money would be taken out and shot." Though T'ang invented his own currency (which, by custom, a governor could do; Rock's objection was to its "softness"), dealt in opium, practised nepotism ("common to politicians all over the world"), and sanctioned wholesale looting, Rock considered him "a gentleman and also a scholar of sorts" compared with his successors. When he died in 1927 there were at least two versions of his death; the American consul used the term "overthrow".[5] His generals formed a committee with Hu Jo-yu as chairman. While Hu, described uncharitably by Rock as a "Szechwan opium sot," was out in the country fighting T'ang's legacy of bandits, he temporarily relinquished his chairmanship to Lung Yun, a general from the Lolo tribe. Lung Yun, which means Dragon Cloud, politicized and gained supporters. When Hu returned the two men vied for power and engaged in a series of battles including one near the American Consulate in mid-June.

Rock happened to be in Yunnanfu at the time. Gingerly picking his way through bodies in the street, he entered the Consulate, found the American representative, Joseph Jacobs, flat out on the carpet in his sitting room, and burst out laughing, leading the Consul to remark acidly, "You are the only cheerful man I've seen today." Jacobs, sensibly fearful of stray gunfire, had the good sense to realize that the fighting was strictly a matter of "personal jealousies of two Chinese generals" and courageously kept the Consulate open though advised by his superiors in Peking that he could shut down.[6] When all the plotting, counter-plotting, and fireworks had subsided, Lung emerged the stronger and chased Hu to Szechwan, where the latter's friend, Liu Wen-hui, a governor-war lord, appointed him border commissioner along the Szechwan-Yunnan border, thus assuring more trouble and evoking Rock's comment that Lung would have been smarter to kill him.

Like his predecessor, Lung printed his own money while stashing away a fortune in sound currency earned by trafficking in opium. Shrewd and wealthy, he held his opponents at bay, maintained an unsteady grip on the governorship, and kept Yunnan semi-independent of Peking. During the Japanese invasion he agreed to throw in his lot with Chiang Kai-shek, but so tenuous was his cooperation that the Generalissimo greatly feared offending him and alienating his support.[*] Lung, a tiny man of gigantic vanity, affected superiority. He took great delight, for example, in accepting dinner invitations from foreign consuls in Yunnanfu and then falling to appear at the last minute, thus causing the foreigners to lose face. The consuls were put in a delicate position; they could scarcely neglect to invite the governor to important functions. One American Consul, Paul Meyer, having been warned about Lung's idiosyncrasies, risked asking him to a formal dinner in honor of the American ambassador who was visiting Yunnanfu. Lung characteristically accepted and then, at the last minute, sent a runner announcing his change of mind. Meyer, playing the game to the end, responded by cancelling the affair. Other examples of Lung's whimsical exercise of

---

[*] Prof. John K. Fairbank, writing in the *New York Times* of Aug. 12, 1971, observed "the enormous Chinese realm has never been a unitary state tightly controlled from the capital. Equal in size to Western Christendom, the Chinese empire was similarly broken up by geography but never became a congeries of separate nations. Maintaining the central power was a constant miracle of political engineering. It required intensive political indoctrination and maintenance of the ruler's presige, but also balancing of central and local interest, cooperation and compromise between the capital and the provinces. Political unity rested on a multitude of local deals such that regional leaders found it easier to acknowledge the central sovereignty than to rebel against it. Neither the emperors of old nor Mao today could control China by simple fiat. Often the central government, like that at Nanking under Chiang-Kai-shek, was acknowledged but bargained with the outlying provinces. Local leaders have often been happy to bow to the center but slow to pay taxes. In short, political-economic autonomy has been a well-established feature of the Chinese landscape." Fairbank used this argument in support of his contention that a workable solution can be found between Peking and Taipei if American forces are withdrawn from Taiwan.

power abound: Barbara Tuchman, in *Stilwell and the American Experience in China,* reports that during the war Lung decreed that "all two-wheeled carts, the common vehicle of the area, must be equipped with rubber tires. He then opened his warehouses to sell the tires he had confiscated during the days of traffic on the Burma Road. After that he passed a new law taxing all carts with rubber tires."

Between displays of temperament and feats of fiscal wizardry, Lung struggled to consolidate his power and chase bandits. In the early days of his rule, Lung's special albatross was Chang Chieh-pa, or Chang the Stammerer, a man who boasted of having murdered 300 people and of eating a raw human heart every day. Chang, a member of the Minchia tribe which inhabited the Tali district, started life as a muletier, but finding it a strenuous and unprofitable career, turned his talents to brigandage. Gradually he accumulated a band numbering around 5,000, who terrorized the countryside, particularly the principal caravan routes. A devout Buddhist with an enviable gift for rationalization, Chang prayed to capture only evil men; he confronted his victims with the words: "You are a rascal or you wouldn't be here," and calmly ordered them to their deaths.

Among his band Chang was a disciplinarian, cutting out the lips of those who lied, killing deserters, and forbidding opium. He had been operating on a large scale since 1922 but T'ang and the generals who succeeded him had been unable to bring him to heel. Familiar as he was with the avenues to political power, Lung became frightened as Chang's strength increased. The bandit plagued the route between Tali and Yunnanfu, keeping headquarters in Tali as a "general." When Lung finally dared to march on him in Tali, in November 1928, Chang kidnapped a priest from the French Catholic Mission and sent word that he would cut his hostage's throat if any troops entered his territory. Eleven days of negotiations followed, during which the priest remained in captivity and the French, the most powerful foreign element in Yunnan, complained to Lung. In the end Lung appointed Chang sub-governor of the Tali district, effectively legitimizing his activities.

Rock, who had been closer than was comfortable to this fiasco, summed up bitterly: "Generals and bandits—the term was synonymous—were as common as lice in the clothes of coolies or even gentlemen in western Yunnan, . . . The more vicious and blood-thirsty a brigand chief, the easier it was for him to become a general, and this is the height of their ambition, . . . especially if there is nothing more to be robbed. At such critical times, when all traffic has practically stopped due to their depredations and the country has been bled white, the hamlets burned, it behooves them to make peace with the military in power. . . . Thus the Government will pay their wages as regular soldiers while their bandit leader becomes the legitimate ruler of the region he has previously ransacked. The poor peasants whose homes have been destroyed but who have no other place to go, their ancestral holding furnishing them a bare living, are then taxed to keep the ex-brigands fed and clothed for a temporary guarantee that their rebuilt homes will not be burned and their very clothes not looted off their backs."

But the metamorphosis was reversible: underpaid soldiers and their non-commissioned officers deserted with their weapons for the more lucrative life of waylaying caravans; peasants, with little to show for their harvests and nothing to do until the next planting, sometimes joined the bands,' particularly before the Chinese New Year when, according to custom, all debts fell due. As long as the gangs remained small and did not pose political threats and left his interests alone, Lung profitted from their existence by charging extortionate rates to traders for armed escorts. And, as if bandits were not enough of a cross for the peasants to bear, the soldier escorts helped themselves "like locusts" to what they wanted in villages, commandeered peasants along the road to tote their bedding, and abused them.

THE PLACE OF OPIUM IN POLITICS

Another source of Lung's income, along with protection money and rubber tires, was opium. The lawbooks bulged with edicts prohibiting its use and cultivation—the first, against smoking, having been issued in 1641—and, since poppy fields cannot be disguised, plants were grown only with the approval and complicity of the provincial authorities. Rock noted that while Chang Chieh-pa banned the use of opium among his men, he forced the peasants in Tali to grow it. Lung also encouraged opium cultivation, and peasants willingly substituted the poppy for food supplying crops. In 1873 E. Colborne Baber estimated that almost one-third of the arable land in Yunnan was given over to opium; a 1906 production figure was given at 4,800,000 kgs. for the province of which over 3,000,000 kgs. remained for home consumption after export.[7] Around the turn of the century, in an attack of mixed guilt and moral indignation, foreign governments led by Britain, which, meanwhile, continued to carry Indian opium into China, pressured the Manchus to suppress smoking, and a new edict was issued to this effect in 1908, while serious endeavors were made to restrict opium production. But the republic and its attendant disorder, the desires of an addicted population, estimated at 2.5% to 5%, and England's withdrawal from the trade in 1917, combined to revive the poppy's popularity as a crop. So intense was its cultivation in the west that hardly a day went by during Rock's passage through agricultural areas when he did not record opium fields in disgust. It bothered him to see the poppies growing in tribal districts because he knew that opium was traditionally a Chinese vice. Tribal leaders forbad its cultivation in principle but were often forced to give in to the excessive demands of the Chinese tax collectors; once the drug became easily available to them, tribesmen picked up the habit.

Chiang Kai-shek's New Life Movement of 1934 in part directed itself against opium abuse, earning Chiang the plaudits of missionaries and the foreign press; one writer claimed that by 1936 land values around Tali dropped one-third and business at the great fairs showed a visible decline.[8] But opium was smoked openly in Yunnanfu in the late 1930s, and commercial activity languished until 11 a.m. when the effects of the previous night's excesses had worn off. "Vinegar Joe" Stilwell commented on Yunnan's "enormous smuggling racket" in opium and Chiang's obvious reluctance to meddle with it at the risk of alienating Lung: the year was 1943.[9] Remarking on the efficacy of opium-suppression campaigns, Cressey said "these occasions have usually been used as opportunities of increasing the revenues for the succession of militarists who have ruled. . . . In such cases the order forbidding further planting had usually carried a footnote saying that those who did cultivate the poppy should be fined so much per *mow*, and the local tax gatherers have proceeded to collect the fine whether the opium was raised or not. Opium yields the highest values which can be obtained from a given area of land and hence is the most attractive crop for taxation."[10]

In one of the less successful ventures of his career, Lung Yun, dissatisfied with the fruits of ordinary taxation, tried to tap the potential for fantastic profits in the export of pure heroin. He sent to Indochina for machinery to refine the raw opium, and the contraption passed through customs as a cement mixer, no doubt leaving in its wake at least one wealthier inspector. Lung installed the machine in a building his government had constructed outside the south gate of Yunnanfu and announced that the government was in the cement business. A French aviator who advised the so-called Yunnanese Air Corps—organized by a French mission in 1923 and, *c.* 1928, comprised of three Briquets, three Caudrons, a captain, a pilot, a mechanic, and a shortage of spare parts—was then retained to fly the heroin out to Indochina. The elaborate operation, however, could not be conducted in absolute secrecy, and word filtered back to the foreign consuls that the cement factory was not what it seemed to be. They used their influence to persuade Lung to give up his plan, but there were no recriminations. Since Lung had no intention of manufacturing cement, the building stood empty until a Greek Chevrolet dealer rented it as a garage for his cars.

The opium poppy was delicate and demanding of tender attention; for best results, harvesting was done at dawn or after sunset. Following collection, the milky latex was left to congeal into a viscous brown liquid which was then carefully cooked and prepared for use. The ingenious Chinese also concocted a cigarette of tobacco, *cannabis,* and opium, ancestor of the present-day pot laced with heroin. For a peasant, the profits from opium were sufficiently greater than those from ordinary crops to merit its cultivation but, due to the imposition of heavy taxes, insufficiently great to elevate him from poverty.

The tragedy of opium, of course, was not its production but its widespread use. Though figures for the 1920s and 1930s var-

ROCK'S EXPEDITION TO MINYA KONKA ASSEMBLED AT THE CHINA INLAND MISSION *National Geographic Society*

ROCK (FIFTH FROM LEFT) WITH ESCORT TO THE KONKALING MOUNTAINS. THE ROCKS WERE PILED UP BY PILGRIMS WHO CIRCUMAMBULATED THE THREE SACRED PEAKS.
National Geographic Society

A CAMPSITE BENEATH THE LIKIANG SNOW RANGE.
National Geographic Society

CAMELS AS PACK ANIMALS INSTEAD OF MULES FOR EXPEDITION
TO THE DRY STEPPES. *Arnold Arboretum of Harvard University*

ROCK'S HOUSEBOAT ON
THE YANGTZE IN SZECHWAN
DURING JOURNEY TO
SUIFU
National Geographic Society

HAIFAN BOY FROM MULI;
WITH ROCK ON JOURNEY
TO KONKALING
MOUNTAINS
National Geographic Society

MINYA KONKA, CHINA'S TALLEST PEAK, 24,900 FEET.
National Geographic Society

THE RULER OF MULI, ENTHRONED IN STATE.
National Geographic Society

ROCK IN THE NATIVE COSTUME OF CHONI. *National Geographic Society*

THE PRINCE OF CHONI AND ROCK. *National Geographic Society*

THE TSUNGKUAN OF YUNGNING AND THE PAVILION ON HIS ISLAND OF NYOROPHU.
National Geographic Society

ied enormously, even the lowest estimates of the numbers of ha-
bituees were alarming. Since the beginning of the century the
production of opium had nearly disappeared in eastern China
but the traffic continued; in the western provinces, notably Yunnan, Szechwan, Kansu, and Kweichow, production, traffic, and
use flourished. In Yunnan opium was cheaper than cigarettes. A
friend of Rock's who worked at the mission maternity home in
Yunnanfu reported that 80% of the expectant mothers had to be
supplied with opium during their confinement. Children smoked
openly in the streets.

Nothing infuriated Rock more than opium. He could not
control the occasional muletiers and soldiers whom he hired on
his expeditions except to forbid them to smoke in his presence.
Toward the Nakhi who were his full-time servants, however, he
maintained unrelenting vigilance. After twenty years he still
sniffed them for telltale odors. Once, suspecting Wang the cook
of misbehavior and blocking his ears to the man's entreaties, he
expelled him without notice, then—cooks experienced in Austrian cuisine being hard to come by—took him back two days later
on Wang's solemn oath never to touch opium. Every ride past an
opium field or night spent in an inn above stupefied soldiers oc-
casioned a rancid Rockism. He was correct in his harsh conclusion that nothing short of total war on opium production and use
could eliminate it from the Chinese landscape. Nothing that happened during the years he spent in China gave him any reason to
believe that such a campaign either would or could be waged.

ROCK BLAMES INDIFFERENCE

Until the Japanese pushed Chiang Kai-shek into Szechwan,
and the Hump and the Burma Road became military factors,
America paid scant attention to Yunnan or the western provinces in general. Unofficially, the United States recognized
French domination in Yunnan which bordered on Indochina,
and the Russian influence in the northwest; the British were observed to have interests in western China, an outgrowth of their
presence in India and Burma. American consuls stationed in
Chungking and Yunnanfu—the latter consular district included

Tibet!—assessed developments within their areas as best they could in reports to which the Legation in Peking and the State Department in Washington assigned secondary importance.

Had Yunnan's anarchy been unique in the Republic of China it might have received a greater proportion of foreign attention. But warlords, opium, bandits, squeeze, corruption, and murder were the norm, and China lived up to the name Sick Man of the Far East applied to it by weary diplomats. Most people who knew China tried, at one time or another, to describe the causes and symptoms of this disease. "The civilization of the Chinese people, who comprise one-fourth of the human race, is being shaken by the impact of the radically different machine civilization of the West with resultant disorder and unrest," essayed Nelson Trusler Johnson, American Minister in Peking *circa* 1930. "An ancient civilization is breaking up, and it will be many years before the work of constructive agencies (for example, the Mass Education Movement) can counterbalance the activity of irresponsible persons to whom a time of disorder presents an opportunity for personal gain at the expense of the community. Such persons are the bandits who roam over the countryside in increasing numbers and the horde of minor militarists who, unhampered by guiding principles, combine with and against one another with readiness and who, as a rule, leave in their wake no more tangible evidences of their presence than further impoverishment of the areas under their control."[11] Johnson's synopsis, in many ways characteristic of the American attitude toward China, was at once generous and self-centered: while graciously accepting on behalf of the West the guilt for China's problems, he denied the Chinese their historical integrity and produced partial explanations.

Rock, meanwhile, from his mountain fastnesses, was not so predisposed to charity. Commenting on a book, *China: a Nation in Evolution* by Paul Monroe, the explorer attacked the author's "silliest and most ridiculous contribution," *i.e.*, his agreement with the Chinese "that their chaotic conditions, militarism, etc., in fact all that ails China, is due to the unequal treaties. I wonder if he knows about the 300 million or more Chinese who do not even know that such treaties exist; they have not the slightest idea what these so-called treaties are all about, who made them, and with whom they were consummated, and still less would they care did they know, for they do not affect the Chinese of the interior. What they are worrying about is, will the parasitic militarists and the other squeezing officials leave them enough to keep them from starving." To attribute the problems of China uniquely to the West was, Rock concluded in August, 1932, "arrogant" and "childish."

In his opinion, the Chinese themselves were largely to blame for the chaos; they were afflicted with *indifference*, a word he repeated tiresomely in his indictment of China. "How easily, by little exertion, these people might be bettered. But why exert yourself? Rather huddle up a little more and just endure a little more discomfort. The only thing that matters is face and conventions. On a certain day they are obliged to don a white turban, burn a few incense sticks along the road in front of their house, and howl and scream, although it is three years since their more or less unbeloved have departed. Should they omit this forced howling, they would lose face. With the farce of their lives they are much more concerned than with the hard facts of reality from which they suffer and which make their lives miserable; but this leaves them indifferent. Lice and fleas and bedbugs will always remain their companions. To these vermin one would almost say they exercise a generosity of live and let live. It is wrong to call it generosity; it is the height of indifference. Why should you go through the exertion of looking carefully to exterminate the beast when you can scratch yourself? Why wash your body when the few rags necessary to cover your nakedness will also cover the dirt?

"Money, although it is almost official counterfeit, is not to be spent but to be buried and hoarded. The sick are allowed to starve, but the dead are buried twice and three times in order to mainly make a show, and if there is no money available they will mortgage their last field and become indebted for several generations to come, but they must give a show to the rest of the village people who know that it is all done on borrowed money.

"I have not been in the interior of Africa or New Guinea or the wild and uncivilized parts of Australia, yet I believe the most comfortless, disreputable, miserable existence is led by the Chinese in the interior of China. One could go on describing this race and their kindred and come to no other conclusion and give no better term of description than the word *indifference* implies in its full meaning. Indifference is responsible for all their plights, from body lice to bandit parasites and officials who squeeze the life blood out of them. Yet they would think [the officials] were fools if they did not take advantage of their . . . position. How quickly they would do it themselves had they the chance . . .

"It is indifference, utter indifference. Indifference to the last degree of possible practice is responsible for the chaos in China. It is the rough element which takes advantage of this indifference to exploit the majority. Militarists are cowardly parasites with little more dare-deviltry who can gather about them elements which are skilled and refined in cruelty and who are ready to oppress their very neighboring villages without the slightest sign of mercy. What is the use of continuing the analysis when the outcome is, and always will be, indifference which breeds selfishness and the latter, in turn, materialism and cruelty."

These views Rock had entered in his diary on November 11, 1928. His eccentricities colored his views: his mistrust of dogma, his mania for personal cleanliness, his intense willfullness, his vanity in being familiar with a region known to so few Westerners. Yet his words should not be dismissed merely on the basis of immoderacy; in some respects he was correct. Certainly he was justified in his irritation with authors whose conclusions rested solely on experience with East Coast treaty ports. Millions of peasants and nomads, scratching out what passed for an existence in the Chinese interior, were indeed ignorant of unequal treaties. In the regions where Rock traveled, he saw few tokens of intercourse with the West, opium and missionaries being the most apparent occidental contributions to the landscape. Yet neither the drug nor the ambassadors of Christianity, even in the most diabolical combinations, could be held accountable for chaos in China.

But in his effort to find a one-word cause for the turmoil, Rock missed the mark. Indifference failed to provide a meaningful interpretation of China's woes, and the failure originated in Rock's own impatience. Measuring others against himself, he could not comprehend the conditions that prevented a man from pulling himself up by the bootstraps, taking a bath, disregarding cultural myths that did him more practical harm than good, and generally putting himself in order. Reared in a culture which glorified the individual, and himself a prototypical self-made man, Rock believed each person had but to exert himself and China's troubles would vanish. The only possible explanation for the absence of this effort was, in his vocabulary, indifference. He did not realize the generations of tradition which had evolved different values and loyalties; centuries of government which, to most people, had meant oppression; loyalty to the extended family rather than the individual; cultural and religious rituals which contributed order and illusion, as well as the promise of an improved afterlife, to an otherwise intolerable reality. If the tax collector bled one white every year, why not turn bandit or grow opium? If the government permitted corruption, why should an official sacrifice himself to honesty? Rock forgot that Western civilization had its share of oppressed people whose sense of futility and desire for dignity expressed themselves in other forms. And so, though he recognized the symptoms, he could not diagnose the Chinese disease.

Other Westerners noting Chinese passivity interpreted it differently. Nathaniel Peffer, an experienced journalist and China-watcher, remarked in *Asia* in December, 1924, that the country's mood was "inert and pessimistic," that there was a widespread "feeling of helplessness." But, unlike Rock, he had faith in the Chinese. "It appears to be a law of Chinese history," he wrote prophetically, "that the Chinese people have to be goaded beyond endurance before they assert themselves. When they finally do, they always leave no doubt that the Chinese people are masters of the Middle Kingdom."[12]

Implicit in Rock's words was an assumption of racial and cultural superiority. Though they might have been reluctant to admit it publicly, very few Westerners thought of the Chinese as equals, extravangant praises of silks and Ming pottery notwithstanding. "The ordinary foreigner in China and Japan," according to Peffer, "lives in his foreign settlement, insulated against irritating attacks of an alien civilization. The environment of the land of his birth is reproduced for him as far as is physically possible." (*Asia*, May, 1924, p. 356). The British, especially, with their Kipling complexes, were often shamelessly outspoken in their prejudices. Brigadier General George Pereira, Rock's acquaintance and a man of much experience in China, was of the opinion that there "is generally a latent feeling of dislike of the foreigner in every Chinese who has any education, chiefly because he realises in his heart of hearts the superiority of the foreigner." He blamed Chinese students for stirring up political trouble and regretted the "zeal of the foreigner for bringing the Chinese up to modern requirements."* Ironically, the same year, 1921, that Pereira wrote those words, Mao Tse-tung, who had had some Western education in Peking, was busy organizing workers in his home province of Hunan. Symbolic acts, beginning with the refusal of Westerners to kow-tow in the presence of the emperor, and overt displays of contempt and smugness—such as signs in the Shanghai foreign settlement reading "No Dogs or Chinese," offended particularly the educated Chinese and polluted relations between East and West.

All Westerners, including American diplomats, were not as universally insensitive to racial questions as General Pereira. Johnson's predecessor as minister, John MacMurray, reminded his superiors in Washington more than once that "the Chinese, in common with other Asiatic people, have particularly since [World War I] been growing more favorably self-conscious, less in awe of the' western peoples, and more determined to assert themselves and resent the assumed superiority of the white races."[13] To all politically minded Chinese, regardless of their orientation, the unequal treaties represented "a stigma of racial inferiority."[14] Awareness of the problem, even by such a lofty official as MacMurray, did not alter the policy or the behavior of the foreigners in China. The West, with the exception of the Soviet Union, which relinquished extrality in 1924 and had its own axe to grind, clung to its special privileges; Westerners persisted to patronize, abuse, underpay, and downgrade their hosts. The

---

* Quoted by Younghusband in *Peking to Lhasa*. One Briton who had no problems of prejudice was the late Bertrand Russell, who once spent a year lecturing on philosophy at Peking and found the Chinese both delightful and quite scrutable. In his book, *The Problem of China* (1922), he compared Western civilization and merchandising mentality very unflatteringly with Chinese values.

Chinese, meanwhile, were also racially biased, though conceding that the West had come up with some wonderful inventions, especially in the line of war material. Chiang Kai-shek, a Christian, thought some white men smelled like raw meat and would order the room aired after they had departed; other Chinese were repelled by the white man's hairiness and odor. In contrast with the Chinese, the white man refused to be insulted. Having tested his potency militarily and economically, he shrugged off the cries of long nose, red-haired barbarian, and foreign devil.

EFFECT OF MISSIONARY ATTITUDES

Missionaries involuntarily contributed to racial tensions by implying through their efforts that the Chinese worshipped inferior gods. The excellent work of medical and teaching missionaries was undermined by their straightforward, hard-sell colleagues who treated the Chinese like sinful children. "As foreigners," wrote Peffer, "they came with open eyes and critical facilities and prejudice against the unfamiliar. Seeing life naked, without the corrective of knowing life at home in the same way, they of course idealized their own countries and adopted an air of superiority."[15] Missionaries accounted for a good deal of the information the Western public received about China, and their interpretations were often dangerously misleading, an occupational hazard. Anxious to raise money for their endeavors, they sent their sponsors letters picturing misled heathens whose only hope for earthly or heavenly salvation lay in conversion. Photographs of winsome Chinese orphans plucked at Christian heart-strings back home as did horror stories of disease, famine, and idolatry. The missionary reports aroused sympathy, particularly in the United States and Britain, but it was the kind of patronizing sympathy that can prove damaging. Christians had been known to raise funds for stray animals, too.

Inside China, in remote and unsettled regions, missionaries frustrated the foreign consuls who tried to look after their welfare and the Chinese officials responsible for their safety under the rules of extrality. Rock recalled how Mrs. Marston, the unfriendly and dimwitted Pentecostal missionary with whom he had stayed in Chingtung, had a habit of wandering off to proselytize

71

among the Lolos without notifying the district magistrate. When her absence was noted, the magistrate would dispatch the chief of police to bring her back; whether or not she consented depended on her conversation with Jesus.

Less difficult individuals than Mrs. Marston, standing upon religious principles, were reluctant to abandon their posts even in case of extreme danger, thus setting themselves up as targets for anti-foreign sentiments and political kidnappings. Chang Chiehpa's capture of the priest from the French mission was only one such incident; in October, 1930, for example, communist forces abducted no less than a dozen missionaries in different parts of the country, demanding ransom for their release, while freelance bandits held others. When missionaries ignored circulars warning them to move to a safe location, the consulate, under pressure from home where a murder or kidnapping of a missionary made for unsavory publicity, could only appeal to harrassed provincial authorities to provide protection. One British consul in Yunnanfu, tired to listening to tales of woe from missionaries who would not budge, tried the tactic of claiming he was an opium-smoking Moslem to chase them from his doorstep, but he, like his peers, had to concern himself with their safety and kept the diplomatic telegraph wires busy. By the mid–1920s, the missionary population had reached a peak of around 12,000, about two-thirds Protestant and one-third Catholic, according to Tuchman. Anti-foreign resentment was simultaneously high, and the Chinese did not distinguish between the foreign businessman and the missionary.

The Chinese, finally, defied facile classification as the befuddled victims of Westernization pictured by Johnson, the innocent heathens characterized by the missionary, of the indifferent, miserable creatures drawn by Rock. Bombarded with reports and opinions, the Western mind, with its inherent love of compartmentalization, was understandably baffled. Only the China expert could follow the intricacies of political warfare in the 1920s; the ordinary man found himself adrift in a sea of undistinguishable names. Nationalists, warlords, and communists battled each other in shifting alliances. After the Kuomintang expelled the communists in 1927, the U.S. State Department floundered about for standard terminology, describing the latter variously as reds, communist-bandits, so-called communists, bandits, and, occasionally, just plain communists. If it was imposible to pinpoint the cause of the chaos or difficult to discriminate between heroes and villains, it was equally impossible to ignore the fact of chaos and to wonder what, if anything, would come of it.

Rock observed all this with a sort of deep internal sigh and concluded that "the only thing constant in China . . . is its instability."[16] He had no illusions that any pretender to power could set things straight. During the early 1920s the fighting in the east affected the western provinces only insofar as the continued absence of effective central government permitted the smaller scale western disorders to proceed unchecked. The roads were unsafe for travel, the villages were looted when it seemed there was nothing left to loot, bandits increased in number, and warlords of varying pedigrees persisted in squeezing the peasants. Rock vented his wrath periodically in diaries and letters, now pitying, now cursing the downtrodden peasants. He lamented opium, bad roads, corruption, soldiers, missionaries, currency irregularities, lice, and Westerners who misjudged what he believed to be the "real" China. Significant omissions from his catalog of complaints were the faithful, "unspoiled" Nakhi tribesmen and the glorious scenery of the mountains.

Countering the bandit menace with money and military escorts, Rock averted any major mishaps. From May to October, 1922, he concentrated his collecting in the general vicinity of Likiang. From September through November he worked the south of the province in the Shweli valley near the Burmese border, hunting the blight-resistant chestnut. In 1923 he followed a similar pattern, remaining in Nguluko until fall. He was diverted for a while by a visit from General Pereira, whom he had met accidently during the winter in Tengyueh and who passed through Likiang in August and stayed with him. Rock headed south in October to the Mekong valley, and to Champuton and the Salween and on to Bangkok at the end of November.

"Never in the world were there such mountains," he wrote; "it is not a long distance to the Salween or Irrawady as the crow flies but, when you are bound down to Mother Earth and you have to crawl like ants over such ranges as are separating these rivers, then you will realize that it takes four days alone to the east bank of the Salween."[17] He collected quantities of seeds and bulbs though exploring in the Mekong valley had been restricted by an epidemic of pneumonic plague. In Bangkok his contract with the National Geographic Society approached an end, and his future was up in the air; but instead of leaving the Orient from Bangkok, which would have been the logical route, he returned to Likiang, to "make a dash" for the territory of Muli in the extreme southwest cornor of Szechwan.

Rock was determined to make his visit before departing, possibly forever, from China. He had tried to make this journey two years earlier and failed because of unusually heavy bandit activity and a warning from the Muli *t'ussu*, or lama king, that a Tibetan tribe, the Hsiang-cheng, had gone on one of its periodic rampages in his territory. He was attracted by the isolation of the place, the difficulty of the trip, "one of the most trying in southwestern China", and the fact that "Europeans who have passed through during the last 100 years can be counted on the fingers of one hand." Ironically, his rivals Forrest and Kingdon Ward accounted for two of those fingers, a coincidence of which Rock was painfully aware and never acknowledged in print, surely a deliberate oversight. It was Forrest who introduced the lovely *Rhododendron muliense* to western gardens.

The former Muli king had died in 1923, and Rock no longer felt bound by his wishes to stay out of the territory. Rather than risk a formal rejection, however, he did not send a runner ahead to ask for permission from the new ruler. He made plans for leaving Likiang in January, 1924, about a month before Chinese New Year, when bandit activity always picked up. He encountered protests from the Likiang district magistrate who did not want the responsibility for the Westerner foisted upon him. Rock insisted, to which the magistrate replied testily that, under the circumstances, he would not furnish an escort. In his article, "Land of the Yellow Lama," in the *National Geographic* Rock wrote: "I sent my card again with the brief remark that I would start the next day at 6 a.m. He made no reply this time, but in the evening there appeared at my village ten Nakhi soldiers fully armed with Austrian guns of the vintage of 1857. These weapons were muzzle-loaders and in woeful condition. Some were tied with string and others were nailed together to keep them from falling apart." Since further appeals to official-dom appeared to be useless, Rock resigned himself to the raga-muffin escort. The caravan—three riding horses, eleven mules, muletiers, personal servants, and soldiers—began its trek in a southwest gale one cold January morning, the dry snow swirling around them.

Muli lay roughly a hundred miles to the north of Likiang, but the terrain was so mountainous and the trails were so poor that the journey took eleven days. Chinese were conspicuously absent in the sparsely settled mountains; Nakhi, Lolo, Lushi, Moso, and Hsifan tribesmen living in hamlets "perched on cliffs like swallows' nests against a wall," ekeing out a living and speaking different languages, reminded Rock of Babel. Contrary to expectations, Rock encountered no bandits; instead, the weather dogged him, and the caravan was assaulted by blizzards and icy temperatures at high altitudes. Crossing the last pass at 15,000 feet, Rock descended on Muli through a steep forest overlooking the Litang river valley. He dispatched a soldier to carry his card to the king "as it would not do to arrive in Muli unheralded," announcing his entrance into the city the following day. At dawn there appeared an unwashed lama bearing an invi-tation from the king for Rock to be his guest. The lama, who was also the king's secretary, led Rock into the town and to his quarters in a comfortable house. Rock waited for his caravan and changed from riding clothes into something more appropri-ate for an audience with a king who, the lama reported, was ea-ger to meet a stranger.

"I doubt whether until that time [the king of Muli] had known of the discovery of America. He did not have the slightest idea of the existence of an ocean, and thought all land to be con-tiguous, for he asked if he could ride horseback from Muli to Washington, and if the latter was near Germany." Robed in a sort of red toga flung over a gold and silver brocade vest, the king, Chote Chaba or Hsiang tz'u Ch'eng cha Pa by name, was a genial fellow of monumental proportions, standing 6'2" and weighing perhaps 300 pounds. "His muscles were weak," Rock observed with disapproval, "as he neither exercises nor works" —the privileges and pitfalls of absolute power. Starved for news of the world beyond his tiny domain, he pestered Rock with questions: Did a president or an emperor rule China? Were the white men still fighting? The latter, in reference to the World War, was posed sensitively with some embarrassment. "Then, the king suddenly held forth his hand, asked me to feel his pulse and tell him how long he was to live! From this he jumped to field glasses, asking if I had a pair with me which would enable him to see through the mountains." From some mysterious source the king had acquired a collection of faded photographs representing Western life: the dining room of the White House, Windsor Castle, Norwegian fjords, a laughing group in a German beer garden, etc. These he submitted one by one while Rock essayed elucidations in terms of Western culture, doubting his success all the time. The question and answer period became tiresome, but Rock was simultaneously fascinated and appalled that a man who wielded so much power could be so formidably child-like. The scene was a study in contrasts for, while the king posed naive questions, the lamas stood by in abject obedience, with bowed heads and folded hands, poised to jump at his command; they were forbidden to sit in his presence. This was the first of many signs which told Rock that though the king might be innocent of affairs outside his bailiwick, he was no stranger to the subtleties of power. Later visits to Muli would convince Rock that this giant of a man was capable of ruthlessness and political cunning.

Delighted by his guest, Chote Chaba entertained with his finest, which was not necessarily without drawbacks for the unsuspecting Rock. Buttered tea, a Tibetan delicacy to which the foreigner is not easily accustomed, was served in porcelain cups set in silver filigree with coral-studded silver covers. The *pièce de résistance*, resting on golden plates and looking very much like a Turkish sweet, proved to be well-aged yak cheese "interspersed with hair." In the face of such overwhelming hospitality, Rock took a deep breath and swallowed. The visit passed agreeably. The king, the lamas, and the Living-Buddha-in-residence posed for photographs which, predictably, found their way into the *National Geographic*. Rock answered more questions, asked some of his own, inspected the lamasery, and observed religious ceremonies. He stayed only about a week but, in that time, he cemented a friendship with the king. Before his departure Chote

Chaba presented him with a load of gifts including a golden bowl, two Buddhas, and a leopard skin; less welcome, but equally well intended, were a dried leg of mutton, yak cheese crawling with maggots, and a wormy ham. Rock responded with three cakes of scented soap for the lama-secretary, silver coins for some of the other lamas and, for the king, a gun and 250 rounds of ammunition. The king begged Rock to return and, though uncertain he could keep his word, the explorer so promised.

From the border kingdom of Muli, the caravan made its way back to Likiang through Yungning and down along the east bank of the Yangtze where, Rock wrote Gilbert Grosvenor, "no white man had ever set foot," arriving at the end of February, in time for a lunar eclipse. "You should have heard the yelling and screaming of the people," he told David Fairchild. "The whole village ran about like mad, beating gongs, drums and howling. They said a huge frog was eating the moon . . . " Low on funds and without an assignment, Rock left Likiang in March of 1924, traveling via Shanghai and Peking, where he became sad and lonely. "My soul still dwells in the great silences among the snow peaks . . . I feel like a deer that has been taken from the wilds and placed in a zoo." Reluctantly, he headed for what he thought of as "civilization."

IV

EXPEDITION INTO KANSU
FOR THE ARNOLD ARBORETUM

When Joseph Rock came to Washington from China in the summer of 1924, he took on the task of identifying some of the 80,000 different plant specimens that he had sent to the Museum of Natural History at the Smithsonian in previous years. Some of these were duplicates, for Rock, like other good botanists, always collected more than one example of a species, which the receiving institution could use as a basis for exchange. Rock took rooms at the Cosmos Club for the duration of his stay, but he did not relish indoor work, especially in the midsummer heat and humidity for which Washington is known, and made plans to repair, as soon as convenient, to the more bearable temperatures of Rockland, Maine, where he planned to write an article on Muli that had been commissioned by the *National Geographic Magazine*.

Rock was duly welcomed as an authority in his field at Smithsonian, where tributes to his scientific achievements and public attention were good for his ego. J. H. Riley, the ornithologist who studied the 1,600 bird skins that were the results of Rock's most recent expedition, discovered three new species of birds from western China, one of which he named *Ithaginis rocki* in honor of the collector. Later he was to find a fourth and also named that for Rock.[1]

"Civilization" disappointed Rock even more than he had expected, a situation that he hastily blamed on civilization, overlooking the possibility that long periods in the Orient had altered his vision of, and responses to, Western life styles. The things which pleased him—abundance of hot running water, good restaurants, cultural and intellectual activity, respect for punctuality —were outweighed by the things that irritated him—traffic, air pollution, fast pace, and a lack of reverence for simple, natural phenomena. Most galling of all was the scarcity and, by contrast to the East, indolence of servants. Accustomed to commanding household lackeys, muletiers, and soldiers who cringed and obeyed his every whim, Rock was non-plussed to discover that it could take half an hour for a room service waiter to deliver a bottle of mineral water to his hotel room; furthermore, the fellow expected gratitude and gratuities. More often than not, however, the poor waiter confronted a medium-sized, red-faced man in an advanced stage of aggravation. The explorer-botanist, raised in the humbleness of servants' quarters in Vienna, was no democrat.

But Rock's visit had started well. Stopping to see old friends in Hawaii, he was surprised and gratified to find himself a local celebrity. The *Honolulu Star Bulletin* of June 7, 1924, published the following agreeable words about him: "As mild a man as ever undertook desperate enterprises; as unassuming a man as ever led forlorn hopes, is Dr. Joseph Rock, visitor today. Many of us remember him as the chap who used to go about the island on leisurely botanical adventures, his eyes shining with the joy of conquest when he had run some shy, rare flower to its lair or successfully finger-printed and identified some unknown plant or tree. This cool and unruffled man of shrubs and science is an explorer as daring in his way as a Peary, a Stefansson or a Scott." "Pohaku" was coming into his own in very respectable company. One wonders, parenthetically, by what accident the editors chose "mild" as a modifier; but Rock did not dispute it. He enjoyed the creature comforts which China had denied him and the flattery of his admirers. Civilization, however, grated on his nerves, and he longed to get back to the Tibetan borderland.

He now found time to get in touch with Charles S. Sargent, director of Harvard's Arnold Arboretum, who, he guessed correctly, would be his ticket back to China. The Arboretum, which had been under Sargent's autocratic jurisdiction since its inception in 1872, had pioneered the large-scale importation of ornamental plants from the Orient to the United States. Sargent himself had made a profitable tour of Japan in 1892, but the Aboretum plant hunter *par excellence* was the Englishman Ernest ("Chinese") Wilson, whom Sargent had lured to his institution with a high salary. Though Wilson was now only 48, he had sustained a nasty injury during an expedition in Szechwan in 1910, which left him with one leg an inch shorter than the other and the feeling he was no longer fit for exploring wilderness areas. He had given up field work in 1919 after botanizing in Taiwan and Korea under much tamer conditions than he had encountered in China. The Arboretum's plant introduction program had been affected also by federal interference. In an effort to prevent invasions of plant pests, the Department of Agriculture had slapped on a series of regulations requiring all alien plant materials to pass through a Washington inspection and, in some cases, through a quarantine and disinfection period. Though this was unquestionably a good idea, Sargent petulantly claimed that Washington inspectors were idiots who would kill all valuable plants and, therefore, it was pointless to continue plant hunting. Secretly, however, he was jealous of the Department's success with collectors like Frank Meyer, David Fairchild, and now Rock. Learning that the latter was momentarily a free agent, Sargent saw his chance to make up for lost time regardless of federal controls.[2]

Following a somewhat laborious correspondence in which money matters were discussed at length, Rock arrived in Boston at the end of July, 1924, and went out to the Arboretum in Jamaica Plain to bargain with Sargent in person. Old Sargent, constitutionally dour and only faintly mellowed in his 80s, had the instinct to recognize a good man when he saw one and the administrative wisdom to trust that man with decisions. Rock, who did not take kindly to orders, appreciated Sargent's deference. The director treated him well, gave him office space and the use of a secretary to make travel arrangements, and wined and dined him at his baronial estate—an honor which the Boston Brahmin seldom accorded to any "employee." Rock charmed Sargent's attractive, widowed daughter, Nathalie Potter, with his witty stories and entertained the staff at the Arboretum, though many of them observed his quick temper and hypersensitivity. Wilson was on hand, and the two collectors swapped adventures; but it was an unequal trade with Rock doing most of the talking. Wilson, always withdrawn, sensed himself displaced by this new Arboretum acquisition and suffered the same feelings of resentment that Kingdon Ward and Forrest had had, though with less reason, since Wilson was already retired with a commendable record. Rock, ebullient and absorbed in his preparations, never noticed Wilson's discomfort. Only after several months in China, when he longed for news and encouragement from America, did it dawn on him that Wilson never answered his letters. He wondered belatedly if he had said something to offend and begged Sargent to explain, but Sargent sensibly ignored the question.

The Arnold Arboretum expedition was to take three years. Rock and Sargent agreed to a sum of $14,000 for the first year, $12,000 for the second, and left the third dependent on the vagaries of Chinese currency and other unforeseeable developments. Of this, $500 per month was earmarked as salary, the balance presumably to cover all field expenses. Before departure, Rock ordered special equipment and instruments, for which the Arboretum paid on the spot. He visited Abercrombie & Fitch in New York to pick out his gear and supervise the packing of his trunks. Harvard's Museum of Comparative Zoology, interested in a bird collection, added $2,000 to Rock's income.

Rock would work areas entirely new to him, specifically the Amne Machin (A-ni-ma-ch'ing shan) and Richthofen (Nan shan) mountain ranges in what is now the province of Tsinghai. Compared with the places he had botanized before, this was indeed *terra incognita*. Sargent's curiosity about the area originated in the fact that, thanks to Wilson, the Arboretum was well stocked with plants from Hupeh and Szechwan and, through exchanges with British and French institutions, had received material from Yunnan as well. Unfortunately many of the species, including some introduced by Wilson, could not withstand the climatic extremes of Massachusetts. Sargent wanted plants from locations farther north, which would be hardy in New England; in fact, in one abortive effort he had sent William Purdom to look for them in Kansu. Since the Arboretum concentrated on ligneous species, Rock was to watch for trees and shrubs, particularly conifers.

Rock was enthusiastic about the itinerary. Ever since his encounter in Likiang with General Pereira, he had been looking for excuses to explore the Amne Machin. Having viewed the range from a distance, Pereira had written: "It towers above everything else in its snow-clad grandeur and must be well over 25,000 feet high as I was at an altitude of 13,000. It looked 30 miles away but was very likely 70 miles off to the southeast."[3] Had he not died of a gastric ulcer a few months after his stay with Rock, Pereira might very well have attempted an exploration of the range; instead he infected Rock with the idea, and a very seductive idea it was because, according to the General's account, the Amne Machin rivalled the Himalayas, and one of its peaks might even top Mr. Everest. Such a discovery, obviously, would be a coup for any explorer. While not professionally interested in the height of mountains, Sargent was captivated by Rock's enthusiasm and aware of the publicity benefits should an Arnold Arboretum expedition find a new Everest. Meanwhile, Rock's primary objectives, the trees and shrubs, would be found in the foothills—or so it was supposed from the size of logs reported to float down the western reaches of the Hwang Ho.

The two men came to terms. Sargent agreed to pay Rock's entire first year salary in a lump sum in advance. Rock bought tents, a folding canvas bathtub, aneroid barometers, cameras, guns, etc., alerted his Nakhi men, cabled his Thai assistant, Boomah, in Hong Kong, of his approach and, in late September, boarded a train for San Francisco to meet his ship and sail for China. He had had quite enough of "civilization."

If Sargent had any qualms about sending Rock into a politically unstable situation, the explorer hastily pooh-poohed them. Inland China and the mountain regions, he claimed reassuringly, were immune to coastal disturbances; bandits, no doubt, would continue to give him a hard time, but by now he knew how to handle them. Shortly before he left America fighting, which had been going on sporadically in the northeastern provinces of China since 1920, broke out anew as three warlords vied for power in Peking: Wu Pei-fu; the upcoming Feng Yu-hsiang, Wu's one-time ally now called the Christian General; and Chang

Tso-lin, the ex-bandit warlord of Manchuria and a Japanese protegee.

The SS *Empress Canada* delivered Rock, his freight, and his ammunition in Shanghai on what he thought was the last day of this war. Unknown to him, Feng had made a deal with Chang and now controlled Peking; Wu, cut off from his armies, had temporarily retired to central China. Shanghai was quiet, Rock reported to Sargent, but scarred. "The surrounding country . . . is in ruins and thousands of people are homeless and have fled to the International Settlement. The International Settlement was barricaded and surrounded by barbed wire entanglements; all wooden bridges connecting the Settlement with Chinese territory had been demolished to prevent Chinese troops from rushing the city."⁴ Communications between the coast and the interior were as bad as ever, and rumors flourished. Rock weighed them unsystematically. He correctly disbelieved one that had T'ang Chi-yao, the Yunnanese governor, murdered by his own men, but worried needlessly about the story that Szechwan —through which he would have to travel—would declare independence. Frankly, he was eager to get out of Shanghai, which reminded him uncomfortably of "civilization." He hurried from one official to another, completing the various formalities and moved on to Hong Kong. There Boomah, upon whom he had counted, was "very impertinent," and as Rock told Sargent, he was in no mood to stand for any nonsense from an Oriental.⁵ He left Boomah behind and continued to Haiphong and Yunnanfu where his Nakhi men presumably waited in readiness. He arrived in mid-November.

FROM YUNNAN TO SZECHWAN THROUGH THE CHINESE BADLANDS

Things in Yunnanfu were not, however, quite in order. Five of his twelve men had been waylaid by bandits on the road from Likiang and had not yet arrived. Furthermore, owing to the commandeering of coolies and mules by the governor for a military expedition into Kiangsi province, both were scarce and prices had soared. Rock was consequently detained in the provincial capital for over a month, trying to organize a caravan while comfortably lodged as a guest of M. S. Myers, the U.S. Consul and an old acquaintance. Rock finally flushed out a muletier who had gone off into the mountains with his animals and agreed on the journey only if Rock paid a stiff rate and provided a written guarantee from the governor that the military would not commandeer the beasts. Rock agreed to the sum, obtained the paper and, having no further word from the missing Nakhis, determined to make for Chengtu in Szechwan over one of China's ancient caravan routes; he left word for the five men to meet him there. The day of his departure, December 13, 1924, dawned brightly, but it was noon before the twenty-six mules, coolies, and heavy military escort were collected and ready to make their way through the throng that had assembled to watch and wish them goodby. Myers accompanied Rock part way through the city, then turned back and waved. "Soon we passed an old temple. Through the fallen wall there greeted me the fat-bellied god, Mi-lei Fu, with a broad grin on his face as though to say: 'What a fool to undertake such a long journey.' "[6]

Rock was not a man to understate a situation—perhaps his sense of drama was in his Hungarian blood. (The Hungarian language, while having many words for *overstatement,* possesses not one for understatement.) This time, however, he guessed correctly. On the fifth day of the expedition, the caravan confronted its first band of robbers, which attacked the pack train and was efficiently dispersed by soldiers. The following day, near the mountain pass of Yakou, the situation became more serious. He wrote Sargent: "We met a large band of brigands. They were on a hill directly in front of us. We retreated immediately on a pine-wooded small hill opposite the brigands, who had numerous dogs. We held the hill, guns in hand, and I watching the brigands with my field glasses."

Rock, having quarreled with dallying soldiers in the morning, now found himself shorthanded; only twelve of his forty soldiers were with him on the hillside. While they observed, the robbers relieved another caravan of seventy loads of cotton and, when cornered by some soldiers who suddenly arrived on the scene, managed to capture the son of the *likin* (road tax) collector. Rock later learned that they knocked out the boy's gold-filled teeth with bayonets and held him for a ransom of rifles, German Mauser pistols, and ammunition.

The balance of Rock's escort finally caught up with him, permitting him to abandon his hillside position and proceed to Tungchuan without mishap, but with considerable wear and tear on his nervous system. Camped in a quiet temple in the otherwise horrid town, he passed Christmas Eve alone writing the most unreassuring of letters to his Harvard sponsor while listening to German carols on his portable phonograph. On Christmas Day the district magistrate appeared to comfort him and assured him that the road ahead was bandit-free; this, apparently, so he would not have to provide a huge escort. The townspeople told other tales: there was famine in northeast Yunnan, and many peasants had turned robber in desperation. Rock, more inclined to believe the people than the official, requested seventy soldiers and was promised them. The morning of his departure, only forty showed up, and he had to settle for what he got. He wrote Sargent:

"It was just after lunch [on December 27th] on a mountain called San Ko when my muleman came running up to me saying that there were brigands behind the caravan. I waited until all the mules had come up and proceeded, but not very far when my men called 'brigands ahead' and at that moment the robbers opened fire on us. One of my soldiers was killed instantly. The other soldiers opened fire on the brigands, and we retreated downhill under constant fire. My handful of soldiers were really brave and kept the brigands back a bit, but they outnumbered us and pursued us. We reached the bottom of the valley, the brigands hard behind us. We had to climb a hill and, once over that hill, . . . I thought we were safe. But I reckoned without the brigands. They followed us to the village of Panyiengai, which they looted and where they captured three soldiers with their guns."

While the Tungchuan soldiers held the bandits at bay, Rock and the mule train advanced to the next village, Yichehsun, where he was delighted to find an additional thirty-five soldiers who had been dispatched for him from Chaotung. Rock ordered them out to reinforce the first lot, but they returned almost immediately saying there were around 600 bandits on the plain and that they did not like the odds. Moreover, they warned Rock that, in the event of an attack, they would try to protect his person but would not guarantee the fate of either the rest of the caravan or the village. Mindful of what had happened at Panyiengai, the villagers were uneasy and unfriendly, blaming the caravan for provoking the bandits and hurriedly looking for places to stash their valuables. Rock retreated to a dilapidated temple in the center of the village and prepared for the worst. He divided some silver among the Nakhis and made up a small pack for himself containing money, ammunition, chocolate, a can of condensed milk, and warm underwear.

It was customary for bandits to fall upon a village at dawn, and as the night dragged on, the soldiers' fears and complaints multiplied. Rock, lying fully dressed upon his cot, fought the temptation to sleep and reassured them. "Never mind the loads and belongings," he said. "If the brigands come, let us escape and leave the loads behind." He promised them good rewards. They resumed their hill posts. "I thought morning would never come," he wrote. Every minute I expected them to rush the place and I listened for the firing to begin." Dawn passed anticlimactically without an attack. Rock called the caravan to order and now the villagers, having survived the harmless night, begged him to stay for a few more days or to leave some soldiers behind to guard them. Rock claimed different priorities: "I said I was not there to protect the village, but myself and the caravan." He departed with both sets of soldiers.

Coolies and unescorted travelers flocked to Rock's caravan along the road and, with the exception of a small skirmish with a few free-lance robbers near the end of the line, the day passed uneventfully. Rock, though exhausted from tension and lack of sleep, remained alert. "Many old beggar women line the trail near the villages," he informed Sargent. "The houses are poor and the people in a pitiable state. No wonder they turn brigand, for they are practically starving." At the end of the stage they descended upon Chiangti, a town of routine filth, flies, and stench, to spend the night. And so it went, a chain of troubled days and unpleasant nights. On December 29, about mid-day, Rock's two sets of soldiers began to fight with each other. "I was paying them well," he told Sargent, but the Chaotung soldiers stopped all the coolies and other travelers who had taken advantage of my escort and tried to extract an escort fee from them. To this the Tungchuan soldiers objected, and the fight began. The former went up a hill and started firing on the other soldiers." Rock, irate, imposed himself bodily between two belligerent officers standing with loaded rifles and bayonets—one of them was bleeding—and achieved a truce by keeping one contigent of soldiers to protect him while sending the other back to guard the loads. Although, thanks to his intervention, no one got hurt seriously, the incident bothered him, representative as it was of China's chaos.

On the last stage of the road to the city of Chaotung, Rock was met by 250 soldiers sent out by the magistrate to rescue him from the bandits by whom, it was presumed, he had been taken prisoner. He therefore made his entry accompanied by a retinue of some 325 soldiers and was greeted by the greatly relieved magistrate.

One may, at this point, reasonably question Rock's travel precautions, which appear to have been a little over-zealous. His preference for luxury items—tinned foods, extra clothing, folding bathtubs, etc.—was a matter of style and comfort, but his fear of bandits may have been exaggerated. E. H. Wilson, reading Rock's accounts back at the Arnold Arboretum, undoubtedly considered the huge armed escorts excessive and their retainer spoiled. Wilson, after all, had botanized some pretty wild country in China with only a handful of native collectors for protection. Rock's caravans were not ordinary for a Westerner, even in the disorganized 1920s and '30s. One of his long-term friends, Henry Corra, traveled all over China and Mongolia in the '30s as a representative of the Texas Company, an oil firm, in the company of a single native interpreter. He used to take sugar, condensed milk, and cocoa; otherwise he lived, and very well, off native fare and the land. As for brigands, he relied upon information from the villagers who always seemed to know which road was safe.[7] The late Peter Fleming and a female journalist traveled from Peking via Lanchow and Sining through Sinkiang and into India with little more than notebooks and knapsacks in 1935.

However, to balance the argument, Marion Duncan, an American missionary in Batang, was partial to larger escorts, more in Rock's style. "Twenty soldiers a day is the average escort for a foreign caravan," he claimed, in *The Yangtze and the Yak*.[8] The traveler's protective measures, then, varied as much with his state of mind as with objective necessities. Rock, in his dramatic style, had completed the first and worst leg of his journey to Kansu. He collapsed for a week's rest. The eve of the New Year found him in bed in an Anglican mission in Chaotung, a martyr to the indignities of dysentery produced, he believed, by mission cooking. He permitted the old year to lapse without comment.

MISERY OF THE PEOPLE IN CHAOTUNG

The man who suffered from civilization received the following impressions of Chaotung and its various denizens and noted them in his diary: "A terrible place; narrow, dingy streets, beggars innumerable, dirt, filth, and the awful odor of stone coal which is burned in every house in open braziers is nauseating. Not a single house has a chimney, and the gases escape out on to the street through open doors and windows. It made me sick and very uncomfortable. One street is in the process of construction; it is the gift of the military official. The street is very narrow, although well paved. The houses are of wood with verandahs along the whole front, a very cheap-looking affair . . .

"The [mission] church looks like a prison from the outside. There seemed to be nobody at home but finally Miss Squire, a nice little old lady, came into the sitting room and bade me welcome. The members of the mission, Anglicans, live in a most stilted atmosphere, continuously apologizing to each other as if they had just met the day before. I cannot stand such nonsensical over-politeness between a few bachelors in this dirty town with no one else around except a few Miao-tzu [Miao tribesmen] . . .

"We had hardly arrived when soldiers came and wanted to commandeer my horses. I went over and told them not to dare to touch my horses. I produced my permit from T'ang Chi-yao and went to the military official who said he did not know I had a permit and, of course, he would not commandeer them. I then saw the magistrate, a nice old man, who whispered that the military official was a very bad fellow."

A steady snowfall added to the misery of the famine-stricken population and delayed Rock five extra days. Every day, confided the magistrate, who at least had the decency to care, half a dozen or more people died from hunger and cold. In religious adherence to the laws of supply and demand, fuel, rice, and other grains had become outrageously expensive, so most people shivered and ate the roots of *Pteridium aquilinum.* This plant is our familiar fiddlehead fern, which has rhizomous roots that can, under dire circumstances, be eaten. Once Rock got over his dysentery, he had time on his hands and ministered to the sick, particularly among the soldiers, who begged for his medicines and attention. He noted a high proportion of superficial infections, ulcers, and skin diseases caused by malnutrition, neglect, and ignorance of the elementary forms of sanitation. Though his patients sometimes showed improvement, he knew the hopelessness of the situation; it was only a matter of time after his departure before most of them would relapse. Bored with the city, disturbed by the misery of its inhabitants, and aggravated by the missionaries who displayed no gratitude for the half-mule load of supplies he had brought for them from Yunnanfu, Rock was relieved when the weather cleared and he could get back on the road to Suifu (Ipin) in Szechwan.

The road from Chaotung to Suifu was miraculously bandit-free. Though Rock traveled during the Chinese New Year—bandit season—he made the eleven-stage journey without any incidents. The second day out of Chaotung he discharged his mulemen because, with the deep snow and subsequent mud, the road had become impossible for the animals. In their place he hired thirty-seven coolies.

"It was only four days from Chinese New Year and people are not willing to spend New Year on the road" he noted. "I arranged with the Mandarin to commandeer coolies, but as soon as it became known all the coolies fled. However, I managed."[9] Each man received a load of "only" sixty catties, almost 80 pounds. They made the border between Yunnan and Szechwan by January 22 and arrived in Suifu three days later.

Suifu, the third largest city in Szechwan, was a metropolis

of approximately 150,000 inhabitants, who paid allegiance and taxes to one General Lu, a warlord who lived off the revenues he collected from the salt wells at Tzu-liuching. Rock, whose mistrust of any Chinese in uniform had become absolute, was disgusted to find the city swarming with most of Lu's 10,000 soldiers, loafing, smoking opium, gambling, and drinking in bibulous tribute to the New Year. Rock stayed with Americans at the Baptist mission and was informed that there was nothing to be gained by trying to befriend Lu; the general was obtusely anti-foreign. Hopes of saving money by securing Lu's cooperation evaporated, and Rock paid, however begrudgingly, the inflated prices for food, fuel, and coolies. He wasted little time in Suifu and went on by river boat to Chengtu, only to find himself in a new muddle.

"Szechwan is indeed probably the most faction-split-up province in China," observed Pereira in 1921,[10] and the passage of years between his travels and Rock's had done nothing to improve the situation. Along with the dubious distinction of political fragmentation, the province claimed other superlatives: the largest population in China and the Chengtu Plain, the principal subregion of the Red Basin. "Nowhere in China, or perhaps nowhere in the world," George B. Cressey wrote in *China's Geographic Foundations*, "is there a more fertile, productive, or thickly populated agricultural area of a similar size." The relationship between population density and arable land is obvious, and both factors in turn affected the political life as warlords vied for the riches of the plain far from coastal interference. Like Yunnan, Szechwan had a mountainous western region sparsely settled by tribespeople; the semi-independent kingdom of Muli was but one example of the distinctly different hill cultures.

Communications between Szechwan and the other Chinese provinces were at worst non-existent, at best primitive. "Animals are rare, carts unknown, railroads but dreams, canals impossible, and the rivers too swift," wrote Cressey. Before the war with Japan, transportation was almost exclusively dependent on China's cheapest commodity, manpower—the coolie. Had communications been better and both national and provincial politics in less of a mess, food and fuel could have been moved from the Red Basin to relieve the 1925 famine in northeast Yunnan; instead, with help potentially a few hundred miles distant, the Yunnanese died like flies. Poor and unequal distribution plus a lack of humanitarian motives among those in power prevented the Chengtu plain from supplying even the needs of its own; Rock saw as many beggars there as he had in the starving villages of Yunnan. The inevitable opium poppy enjoyed its usual favor and consumed food producing acres. The late growing season by comparison with Yunnan initially fooled Rock into thinking that opium was under control in Szechwan, but a few weeks cured him of this illusion. The bandit hordes that ravaged Yunnan had their Szechwanese counterparts.

Chengtu, nevertheless, made a good impression on Rock. The three bridges mentioned by Marco Polo still functioned. On Feb. 15, 1925, he noted in his diary: "There is feverish activity in the streets of this city . . . General Yang [the current power] is widening all the streets by forty feet and is even putting in sidewalks. This he started last August. He had, at first, great opposition, but anyone who refused to obey his orders he threatened to tear down his house and charge him for doing it. Many people lost all their property by the widening of the streets and some committed suicide, as they had been ruined. I never saw such wholesale improvements instigated in a Chinese town by a Chinese governor. The streets are really pleasant, the value of the property has been considerably enhanced on both sides of the streets. This the people now realize, and there is less opposition to Governor Yang's doings."

Rock was ill with flu for the first eight days of his visit but he submitted himself to the care of medical personnel at the Canadian Methodist hospital and recovered nicely. The city was well stocked with foreign goods—medicines, Kodak supplies, leatherwear—which he needed for the journey to the mountains. He browsed, intrigued by the Chinese medicines for which the province was famous, visited West China Union University, a Protestant supported institution, and befriended many of the Westerners on its faculty. Their efforts to educate young Chinese in useful ways genuinely impressed the normally cynical Rock and, as he discovered, sometimes produced bizarre situations. "There is a dissecting room," he remarked. "The corpses are executed robbers furnished by the governor. They are usually shot, but of late, in order that the bodies should not be too mutilated, they chloroform them to death."

Upon Rock's arrival the illustrious General Yang Sen was off on a military expedition in Mienchow, five days to the north, in an effort to subdue a rival. Since bandits were said to be plentiful north of Chengtu, Rock wanted to apply to him directly for an escort and so awaited his return, and return he did, in triumph and corresponding good humor. The news for Rock, however, was not necessarily good. Ten thousand troops of the defeated Mienchow warlord, after sacking the town on their retreat to leave as little as possible for the victors, had been turned loose and now roamed the hills as bandits making the road between Chengtu and Mienchow a death-trap. Rock heard the sorry details from General Yang himself, a man whose modesty and modernity struck him. Yang admitted ruefully that he exercised only tenuous control outside the immediate Chengtu vicinity and advised Rock of safer roads. By this time, however, Rock's concern with bandits was obsessive, and he determined to wait "till the situation clears . . . . A Chinese proverb says 'when you are in a hurry, sit down,' and this is what I will have to do . . . "[11] While 'sitting down' he got better acquainted with Yang.

The general, a Wu P'ei-fu partisan who operated under the title of Military Rehabilitation Commissioner, was a busy man, desperately trying to consolidate his power. (He later proclaimed loyalty to Chiang Kai-shek and, in payment for services rendered, was awarded the governorship of Kweichow Province in 1945.) Having succeeded, temporarily at least, in Mienchow, he struck south in an attempt to oust General Lu of Suifu. Instead of marching on the city, Yang's troops converged on the Tzuliuching salt wells and tried to collect the revenues; Lu's soldiers marched out to meet them and, at the time of Rock's departure, the two armies, approximately equal in strength, faced each other undecisively. Lu, meanwhile, appealed to warlords in Kiating and Chungking to combine with him to drive out Yang; the defeated Mienchow general was also eager to participate in this plot. Such activity, obviously, was what Pereira had meant by "faction-split-up."

Rock sympathized with Yang, partly because the general flattered him with official invitations and a banquet, partly because, instead of inflating himself with vanity like so many other administrators of Rock's acquaintance, Yang actually made improvements, albeit by drastic tactics. But, like virtually every politician in China, he had opponents by the score and complained bitterly about independent generals and their legions. " . . . It seems," Rock mused, "that everyone wants the other fellow to reduce his soldiers while he recruits new ones every day, and thus there is an increase in soldiers instead of [a] reduction."[12]

Altogether, Rock enjoyed the attentions of Yang Sen, the company of Westerners at the University, and Chengtu as a whole. He rested his nerves which had deen overtaxed on the road north. However, when the situation on the trail to Mienchow did not improve after a month, and spring could be felt in the air, his return to the road became imperative for botanical reasons. Chinese proverbs or not, further delay did not seem expedient. On March 16th he finally resolved to make a run for Mienchow via a narrow mountain trail favored by Yang. He left Chengtu with his twelve Nakhis, the delinquent five having caught up at last; seventeen muletiers, and an escort of 140 of Yang's handpicked soldiers. "Often when I looked back from a hill my train was over half a mile long . . . . We were quiet a formidable party."[13] He later learned that, on that very day, two English missionaries had fallen prey to brigands. This he dutifully conveyed to his Harvard sponsor lest his bodyguard and expenses be considered extravagant.

Chengtu to Mienchow was only five stages. The main road was nearly empty while, on the alternate route, travelers took advantage of Rock's caravan to make the trip. "What prevents the robbers from leaving the main road for the small road I cannot grasp," he wrote in his diary. Ghastly reminders of Yang's military escapade littered the trail: mangled and dead soldiers, half-dead stragglers. Rock received a fresh escort at each town; on the last and most dangerous stage to Mienchow he commanded 190 soldiers. "This all makes traveling fearfully expensive," he explained to Sargent. "Unless one pays these soldiers they are apt to turn one over to the brigands . . . . I shall be glad to get into Kansu and the wilds which is the safest place one can go to. Where there are only tribe people and no Chinese, there is nothing to rob and no ex-soldier-brigands [who] have their haunts on rich plains near large towns where there is a good deal of traffic and hence much opportunity to loot and rob. I have always found the wildest regions the safest."[14]

MOSLEMS AND TIBETANS IN THE LAND OF THE LIVING BUDDHA

Rock planned to tackle the Amne Machin range first and had so agreed with Sargent. His next geographic objective, therefore, was the semi-independent principality of Choni in southwest Kansu recommended by Gen. Pereira. Through Choni's ruler, who was reported to be fascinated with foreign objects, Rock hoped to obtain assistance for his expedition to the mountains. Rock's route took him through Chungpa to Pikou, the first town beyond the Kansu border, then through Kaichou, Minchou, and, finally, to Choni where he arrived on April 21. Though much less harrowing than the passage through northeast Yunnan, or between Chengtu and Mienchow, the journey was not problem-free, nor was the province of Kansu the peaceable kingdom of his wishful thinking. The caravan reached Choni intact having been bothered along its course by unpleasantnesses too insignificant to chronicle here but cumulatively discouraging.

Rock, of course, detailed them in his diaries. "There seems to be something in the air," he declared. By the time he reached Choni, the situation was unfortunately clear to him; he had exchanged one set of troubles for another. Gone were the bandit hordes of Yunnan and Szechwan. In their place was an old hostility between the Moslems and Tibetan tribes which threatened to erupt into full-scale warfare at any moment. Province to province communications were so poor that Rock had heard absolutely nothing of these conditions from either Chinese officials or Westerners in Szechwan and Yunnan. Once in Kansu, the more he learned, the more apprehensive he became about the success of his expedition. All the objects of his botanical and geographical interest lay beyond hostile areas and could only be reached by traversing them. Finding Choni inexpensive and its prince accommodating, Rock called his party to a temporary halt and resolved to wait and see if things would settle down. Spring was late in Kansu and would be still later in the mountains.

Though there was a Chinese governor in the provincial capital of Lanchow (Lanchou), he exercised practically no control in western and southwestern parts of the province where very few Chinese lived.* In the inhospitable regions of mountains and barren steppes lived Mongol and Tibetan tribes, many of them nomadic, some of them warlike, and around two million Moslems. Eric Teichman, a British consular official who traveled in Kansu during the first World War, thought highly of the Moslems. "The superiority of the Moslems over the Chinese in regard to housing, food, personal cleanliness, and general standard of living is marked, . . . " he observed. "Though the magistrates in the Mohammedan districts are Chinese, they usually refrain from any interference with the Mohammedans, who appear to manage their own affairs."[15]

The Moslems long had been dominated by the powerful Ma clan. (Ma is the Chinese equivalent of Muhammed.) When Rock entered Kansu in 1925 the prevailing power in the non-Chinese areas of the province was General Ma Chi-fu, a coarse 6'3" man who made his headquarters in Sining. Moslems, with their long-standing hatred of Chinese and memories of their losses in the rebellion against the Manchus in the 19th century, were loyal to Ma for obvious reasons. The Tibetans, split into factions and outnumbered by the Moslems, had submitted grudgingly to his authority. General Ma and his armies had consistently taxed, fined, and squeezed the tribes and lamaseries in his domain of every ounce of silver he could extract but it appeared, now, that he had finally pushed the Tibetans too far.

The current argument focussed on Labrang, the largest and

---

* Owen Lattimore discussed Kansu in a chapter called "Oases and deserts of Central Asia" and explained the difficulties of the Chinese in controlling the northwest corridor region. In conclusion he summarized: "As a matter of fact, [the Chinese] had no real power of any kind—economic, political, or military—but ruled on sufferance as the least obnoxious people in sight, insulating the oases from the steppes, the townsmen from the peasants and tribesmen, the Moslems from the infidels, and supervising whatever circulation of trade was convenient." p. 188.

most influential lamasery in western China, housing almost 5,000 lama priests and some eighty minor incarnations. Labrang achieved its importance as the residence of a Living Buddha who ranked only behind the Dalai Lama and the Panchen Lama in the lamastic hierarchy. That this incarnation was a youth of 12 years ("a cute little youngster," in Rock's words) whose father had once been a bandit subtracted not one iota from his sanctity in the eyes of his followers. Pilgrims flocked to Labrang by the thousands to seek his blessing; because of his presence the city became a commercial center for both Mohammedan and Tibetan tradesmen. Mindful of this source of revenue, General Ma milked it for eight years and, lately, had dispatched soldiers under the command of one of his brothers. The brother, miscalculating the Tibetan temperament, levied exhorbitant new fines upon the lamasery, payable in rifles and silver. Early in 1925, as a protest to the most recent demands, the Living Buddha's father, who was also the Acting Regent, calmly bundled up his sacred son and, with a retinue of lamas, deserted Labrang. The party took refuge in Angkor Gomba, a mountain temple in Choni territory, which could only be reached by pulling oneself up on iron chains over sheer cliffs.[16] Deprived of its spiritual attraction, Labrang became a ghost town; the markets emptied, and the Moslems—both the traders in Labrang and Ma in Sining—suddenly found themselves dispossessed of a substantial income. The outraged general threatened to bring the Living Buddha back to Labrang by force; the Living Buddha, as expressed by his father the Regent, refused to budge until Ma rescinded his order for money and withdrew every single soldier from Labrang. He stayed in the drafty mountain temple, prepared to move at a moment's notice. His followers always knew his whereabouts, and the pilgrims who would otherwise have gone to Labrang circled his make-shift headquarters, prostrating themselves, whirling their prayer wheels, and chanting the eternal *Om mani padme hum*. The Living Buddha, meanwhile, dispatched a call to arms to all Tibetans on the Yellow River, asking them to assemble for a showdown on the fifteenth of the second fourth moon, or mid-June 1925.

The plight of the Living Buddha of Labrang dramatized the Tibetans loathing for the Moslems and encouraged a general revolt against Moslem authority among scattered tribesmen. The Ngoloks, for example, a nomadic tribe with a reputation for fierceness, had no vested interest in the geographic location of the Living Buddha but wanted revenge for the grass and water tax on every head of cattle which Ma had forced upon them.[17] Alone they were impotent, but in league with other incensed Tibetans they hoped to oust Moslem tax collectors.

The Chinese had been content with the role of observer. Though they had no love for the Tibetans, they did have an interest in ridding themselves of the powerful Ma who dominated vast areas to the west and collected an enviable income therefrom. It was greatly feared, by both Moslems and Westerners in Kansu, that should fighting break out, the Chinese would align themselves with the Tibetans, thereby raising the stakes and the specter of a new massacre of Moslems. Anxious to see Ma's grip broken, particularly at someone else's expense, the Chinese played coy politics and encouraged the Tibetan revolt. The governor in Lanchow and a Chinese general advised the Living Buddha to wage his war and promised to support him.[18]

These are the broad outlines of the tense situation in Kansu in the spring and early summer of 1925, but there were complications in both Tibetan and Moslem camps. The Tibetans were a motley assortment of independent tribes, bound together only by their hatred of Moslems. They had often fought bitterly among themselves. Such alliances make for loose organization and inefficient war machinery. Ma Chi-fu, by contrast, commanded a large army of loyal and well trained troops. But he was unequal to the challenge of the Chinese should they cast their lot with the Tibetans, and he feared this could happen.

One observes somewhat grimly that, while this drama was building in the northwest, another scene, with greater international implications, unfolded in the treaty ports, beginning with the *Wu-san ts'an-an*, the "May 30th Atrocious Incident." Violence erupted in Shanghai when British police fired into a mob of students demonstrating their support of striking textile workers, killing twelve and wounding seventeen of them. Chinese hatred of the foreigner reached a frenzy unmatched since the Boxer rebellion. Though the aliens with their unequal treaties and special prerogatives were sources of legitimate grievances, they also pro-

97

vided convenient scapegoats for opportunistic political cliques during disturbed times. Much as the Arab nations blame all their woes on Israel, so the Chinese—no doubt with more justification —cited the foreigners and the treaties. It was during the aftermath of the Shanghai incident, in July, 1925, that Chiang Kaishek and his Kuomintang forces saw their chance and proclaimed a government in Canton, vowing to abolish the offensive agreements.

The Moslems and Tibetans of Kansu did not have the slightest notion of developments on the coast, nor had they ever. Erich Teichman remarked in 1917, in *Travels in Northwest China,* for example, that "we met a military officer . . . . [in Kansu] who calmly informed us with little apparent interest that China and Japan were at war. This news afterwards turned out to be a local echo of the crisis existing between the two countries two months previously . . . "[19] The foreign consuls in the treaty ports (the Soviets excepted), absorbed in their own difficulties, paid scant attention to interior disturbances. In the confused years of 1925 to 1928, the whole of China seemed orchestrated in a great military fugue of which the Tibetan-Moslem conflict was but one of many variations.

Nothing went according to schedule for Rock. Every time he escaped one set of problems, he encountered another. The responsibility for the caravan, especially for the twelve Nakhis, and the constant proximity to real or imagined danger began to exact a terrible toll on his nerves. He thought wishfully about death and wrote sometimes hair-raising, sometimes morose letters to Sargent in explanation of the many delays. The latter, alarmed and not anxious to have a dead explorer on his conscience, made it plain in his replies that he valued the man more than botanical novelties and that Rock could cancel the expedition if necessary. But Rock, drawn like a moth to the flame, insisted upon staying while continuing his chronicle of complaints. He established residence in little Choni. There he had clean quarters, a friendly host, and political neutrality. Furthermore, should he so desire, two Pentecostal missionaries, Messrs. Derk and Hulton, were on hand for English language conversation.

In the face of all advice to the contrary, Rock remained intent upon approaching the Amne Machin from their western side. In April and May 1925, while awaiting the outcome of the Tibetan-Moslem feud, he made two visits to the Living Buddha of Labrang in his mountain refuge and, on the strength of the Choni prince's recommendation, secured from him (or, rather, from the Regent) introductory letters to Tibetan tribes in the Amne Machin region. Returning to Choni after the second visit, Rock discovered that the Moslems in Taochow (Lintan) thought him an agent of the Tibetans and Chinese arrived to furnish arms. With Rock visiting the Labrang Buddha, it is easy to understand how the story originated. Westerners in that part of the world were usually either missionaries or government representatives; Rock, being neither aroused suspicion. Such gossip was dangerous, and Rock dispatched an agitated letter to Governor Lu Hung-t'ao in Lanchow demanding that he take measures to stop the rumors at once; the governor promised to pass such instructions to the magistrate in Taochow.[20] The incident finally blew over insignificantly but contributed to Rock's uneasiness. The journey north to Kansu had been made bearable by the thought of peaceful exploring ahead. These hopes frustrated, Rock reacted to the tension. Never easy going, he now flew into rages over trivial mishaps, and his moods fluctuated dramatically from day to day. He was no good at waiting, and the wretched Tibetan-Moslem business refused to resolve itself at his convenience.

To fill up time pending a decision, Rock made three trips during the summer of 1925 to the valleys between the Min Shan and the Amne Machin, an area ostensibly under the control of the Choni prince. The unruly Tibetan tribesmen, called Tebbus or Thewus, who inhabited the valley struck Rock as undersized and moronic, and he did not trust any of them. But the valley supported a rich forest flora and was, botanically speaking, virgin territory, giving him that satisfying "first-white-man" sensation.* Trying to keep a safe distance from the Tebbus, which he managed on all but one occasion when five armed bandits attacked the party and shot his Choni guide in the arm, Rock and

---

* William Purdom, an Englishman attempting to collect there for the Arnold Arboretum in 1911, got caught in the crossfire between "Upper" and "Lower" Tebbus and withdrew quickly.

the Nakhis did some worthwhile collecting, the first of the expedition. The activity soothed Rock's nerves.

Meanwhile, the situation between Moslems and Tibetans deteriorated. Returning from the first foray into Tebbu country in mid-June, Rock arranged for yaks, horses, guides, saddles, odds and ends as gifts for tribespeople, and was poised to make for the mountains when fighting broke out between the two belligerents. The Tibetans descended upon Labrang as planned and drove out the Moslem soldiers; those that they captured were strung up by the arms and disemboweled, Rock told Sargent. Ma Chi-fu went on the rampage to avenge the attack and moved troops toward Labrang. Moslems cut the wires between Lanchow and Peking. Three thousand Moslem troops recovered Labrang within a week after its fall, mowing down the Tibetans with machine guns. The lamas all fled, and the band of 40,000 poorly equipped Tibetans scattered in the hills near the lamasery while the Living Buddha and the Regent went to Lanchow to see the Chinese governor and claim the assistance they had been promised. In all the confusion, bandits flourished. Choni was surrounded by Moslems except to the west, Tebbu country, so Rock again postponed his plans and returned whence he had come promising, however, that "I still expect to reach the Amne Machin this year for early fall or late summer collecting."

When he returned to Choni from the valley the second time, a month later, Ma still controlled Labrang, but the Tibetans had taken their revenge by ransacking the lamasery at Heitso, torturing and killing all Mohammedans and burning Moslemowned houses as well as those belonging to lamas who had rented to Moslems. General Ma's response was to offer a reward of three dollars silver for every Tibetan head—man, woman, or child—which his troops brought in. Thus encouraged, the soldiers rode out in the grassy steppes west of Labrang, murdered at random, tied the heads to their saddles, and galloped back to claim their cash. Nomadic Tibetans, poorly armed, had little chance to fight back.

The continuing spectacle of wasted life, materials, and energy which could have been usefully deployed in peace made Rock more cynical than ever about China. "I think the people in this part of the world are all thieves, liars, and robbers and murderers, . . . " he wrote in his diary. "The saddest thing is that they are being supplied indiscriminately with arms from Russia. They are smuggled through Mongolia, and traders, by crossing the Ordos [Inner Mongolian desert] where no Chinese keep guard, can get them through. They sell them at a tremendous profit; old German guns from the 1860s sell for as much as 240 taels apiece. The Moslems themselves sell arms [to Tibetans] instead of confiscating any . . . these people may possess."

Despite the entreaties of the Living Buddha of Labrang, the Chinese had left the Tibetans in a lurch. Ma Chi-fu's success disappointed but did not endanger the Chinese, and the Kansu warlords were unwilling to engage him. It happened, however, that the Panchen Lama—the second ranking official in the lamaistic hierarchy—was in Peking and was fully informed of events in Kansu. When his appeals for assistance failed to elicit a positive response, the Lama was said to have remarked that if the Chinese could not, or would not, help the Tibetens, the latter would ask the English to take Tibet, and thereby planted a bee in the Chinese bonnet.*

The temporary passivity of the Chinese, however, gave Rock hope that the way to the Amne Machin would open. Once more he made preparations, and once more he changed his mind at the last minute for fear that he would never get out alive. Finally, weary of indecision, he gave up the plan altogether for 1925 and determined that, after one more excursion to finish collecting among the Tebbus, he would risk the road via Sining

---

* Lattimore observed that British interest in Tibet was never territorial but, rather, that Tibet was useful to British domination in India. "The prestige indispensable to the rule of the British over India demands that their subjects shall not be allowed to see on any horizon the rise of a power even remotely comparable to that of the British . . . Consequently, British policy in Tibet aims not even at extensive exploration or the exploitation of possible mineral wealth but simply at keeping Tibet inert under the unchanging rule of potentates who will look to the British for support against any encroachment by the Chinese or others." Accordingly, it is unlikely that the British would have "taken" Tibet in any administrative sense; moreover, it is equally unlikely that the Tibetans or the Panchen Lama wanted to be taken. If he did, indeed, make the rumored threat in Peking, he only played upon the deep-seated Chinese hatred of Britain.

to the Nan shan—this in the face of specific advice of the governor in Lanchow against such a journey.[21] Rock reasoned, astutely, as events proved, that as long as the Chinese stayed out of action he would be safest traveling with Moslem escorts in areas firmly under Ma Chi-fu's thumb. This logic, obviously, was unpalatable to the Chinese governor, and Rock did not press his point when he arrived in Lanchow ("the dirtiest Chinese capital I've ever seen") in mid-August, 1925, to request the necessary papers for passage to Lake Kokonor and the Nan Shan.*

Governor Lu Hung-t'ao was a middle-aged stroke victim whose mild and polite protests could easily be ignored; his subordinate, Li Ch'ang-ch'ing, seemed to be the real power. A polished and crafty politician, Li was indifferent to Rock's safety and obligingly produced the documents.

The caravan proceeded from Lanchow to Ma Chi-fu's bailiwick, Sining, without a single incident, arriving at the beginning of September. According to his habit, Rock headed straight for the General himself, bearing a .45 Colt automatic and cartridges as his gift.[22] "The yamen needs a thorough cleaning," he noted squeamishly but, much to his surprise, he came away after a few hours with more than he had hoped. Not only would Ma arrange for mules, escort, and safe conduct to the Kokonor, but he also affably agreed to help Rock get to the Amne Machin in 1926. The meeting between the massive Moslem who paid for Tibetan heads and the choleric little Viennese went off splendidly, testimony to Rock's diplomacy. "We parted best of friends," he recorded triumphantly. By September 15 he was off for Lake Kokonor and the mountains.

---

* More recently Lanchow has been described as "the key communications junction of China's vast but only partially developed Northwest." Highways, railways, and airlines converge there. (C. T. Hu, *China*, p. 52.)

V

AN EXPLORER'S DIARY AS A REVELATION OF CHARACTER

Rock canters across the dusty, wind-swept northwest where the steppes extend with boring regularity. Occasional clusters of nomad yurts, or rounded tents, guarded by mastiffs, relieve the monotony. There is little vegetation, and the prevailing colors are shades of brown and gray, more reminiscent of moonscape than landscape. At night violent gales, which the Chinese call the "black winds," test every fiber of Rock's Abercrombie & Fitch tent and, though the product lives up to the manufacturer's promises, its occupant cannot sleep for the howling outside. He wonders if the incessant wind will drive him mad.

As company Rock has his Nakhi servants, for whom the landscape is equally strange and who are homesick for their native Likiang. They look to him for reassurance, protection, and for the bonus he has promised them when, and if, they all return safely to Yunnan. In his role of guardian and benefactor, Rock holds a godlike power over them; he cannot, therefore, share his doubts and fears with them. For all his admiration—sometimes envy—of their gentle simplicity, they are to him primitive, uncultured, sub-human. As for the Moslem escort assigned to him by his "friend" Ma Chi-fu, Rock loathes them. The "wily wretches" reek of garlic and respond sullenly to his orders. More than once he has caught them eyeing him with frank hatred and he knows it is only the threat of General Ma's machine guns that prevents them from murdering him and making off with his silver. He likens them to "human vultures" and must constantly assert his authority to keep them in line. He cannot snatch a moment's relaxation in this god forsaken wasteland. At the end of a day, when the stars shine and the winds congregate to rage, the Moslems sit joking around their yak dung fires while the Nakhis huddle in conversation. The explorer, accounting for the day in his diary, senses himself utterly and uneasily alone. This is the life he had chosen.

As the popularity of the *National Geographic Magazine* testifies, people have always been fascinated by the feats of explorers and by the psychological qualities of the men themselves. that classic response of mountain climbers ("because it's there") to the inevitable "why?" satisfies, in the fullness of its implications, to explain the men who climb Everest or look for the North Pole or walk on the moon. It does not, however, provide any enlightenment in the case of Joseph Rock, an oddball who selected to spend most of his adult life removed in time and space from his native culture, alienated from people who shared his values and understanding of reality. The hostile steppes of the Kokonor plateau are only a convenient metaphor. In the lamasery at Choni, in the gorges of the Yangtze or—though considerably more comfortably—in a well-appointed house in Likiang or Yunnanfu, Rock is always the consummate loner, taking extreme measures to secure his isolation and then, having achieved it, ill at ease with it. He scorns the company of Westerners, and prolonged intercourse with his peers almost inevitably leads to quarrels. In the end, he turns inward and confesses, with chilling candor in his diary, "I love nobody."

However accidental his geographic perambulations may have seemed, Rock's psychological retreat, expressed by his choice of China, progressed deliberately, according to patterns set in his Viennese childhood. In the absence of lovers and confidants, he became a diarist, and it is in reading these volumes that one begins to touch the essential Joseph Rock. The public images he promoted—the charming witty explorer of dinner parties, the fearless adventurer, the dedicated scholar—fail to capture him singularly or in combination; they are paper dolls.

As he produced his diaries over the years he spent in China, Rock never really made up his mind whether he considered them public property or not; both style and content reflect his indecision. There is much self-conscious writing such as afflicts all but the absolutely honest, private, or happily skilled diarist. Synthetic sentiments often compliment belabored prose, as though Posterity had whispered in his ear, "Come on, now, show them what a good fellow you really were," or "Let them know how much you suffered!" But there were also days when Posterity left him in peace, and agitated, weary, abject, or triumphant, he wrote whatever came into his head.

The idea that the diaries might find their way into the public domain was certainly justified. Rock had explored little or unknown areas, and his records abounded with geographic and cultural data. Even while Rock was still active, the U.S. Central Intellignece Aency, with his permission, sent a secretary with security clearance to Hawaii to type copies of what were considered pertinent volumes. Every day that he traveled he made meticulous notes of compass bearings, landmarks, distances, altitudes, physical features, geological formations, botanical and zoological phenomena agricultural, patterns, and cultural oddities; the information was encyclopedic in quantity, succinct in quality. He also recorded vignettes of travel, encounters with bandits, meetings with chieftans, descriptions of villages, or whatever happened to distinguish one day from another.

Less regularly, but still frequently, Rock's written musings became quite personal, and one can include under "personal" his diatribes on Chinese politics as well as his metaphysical ponderings because both are equally subjective. His reliance on the diaries as the principal outlet for his emotions is even clearer in the pages he wrote when he stayed in one place for a long time, such as Choni in the winter of 1926-27, or in Likiang or Yunnanfu during the 1930s. With no geographical or botanical data to record, he made only random entries, but almost exclusively in times of physical or mental stress. Mindful of potential biographers, he sometimes withheld illuminating details, but he let his feelings spill over the pages. He was a proud man, incapable of disclosing his private longing and terrors to others. He considered emotional confidences undignified and weak. Yet, like all men, he craved understanding, and perhaps that is why the confessions of emotional torment in his diaries sometimes seem almost shameless.

When reading personal diaries, it is healthy to be a little suspicious of certain items in the interest of truth. Such skepticism must be exercised in Rock's case because, as noted in Ch. II, it is now apparent that he lied about his affiliations with the University of Vienna and the College of Hawaii and, years later, to Alvin Chock, who interviewed him specifically to secure biographical data. Allowing for human error, there is no reason to doubt the accuracy or veracity of his scientific observations, nor does he appear to have fabricated any of the dangerous situations in which he found himself. However, as a high strung man given to self-aggrandizement, he embellished situations. I am not, therefore, certain that one can accept all details at face value, such as numbers of bandits or his statements regarding roads being "closed" to travelers. I believe, however, that he was faithful to the essence of the story: there were many bandits, and the roads were indeed hazardous. Discrepancies between diary entries, letters, and his unpublished manuscripts point to a tendency toward excessiveness—a question of degree only and a familiar human foible.

In the back of his last little blue address book Rock scribbled the words: "A bachelor is a man who didn't make the same mistake once!"—a wry comment on his own condition.[1] He never married, probably never proposed marriage and, to the best of anyone's knowledge, never had any intimate relations with women. That he remained single could easily have been a practical decision; singularity was an elementary condition of the freedom he prized so highly and it gave him unlimited mobility. More than one explorer, including E. H. Wilson, had left behind a pining, dissatisfied wife. Logic, then, dictated bachelorhood, but something beyond logic governed Rock's sexual attitudes. Not only did he remain single; he was seemingly celibate, and celibacy is not a logical decision.[2]

Rock's unusual sex life or, rather, the lack of it, gave rise to numerous conjectures among his acquaintances. Some even suspected him of homosexuality, a speculation derived not on the basis of evidence but by default in the absence of conventional liaisons. There is not, however, one substantial indication to support any of the guesses, and one must assume, therefore, that Rock was incredibly discreet or that, in fact, he abstained from sexual experience altogether. He claimed to have renounced sexual desire, but the emotional dynamics of such a decision are mysterious, and there is no certainty that even Rock understood them. Regardless of the causes of his celibacy, however, the absence of heirs bothered him—a species of guilt inspired by youthful religious indoctrination, formal proceedings being easier to shed than moral conditioning. In his mind, childlessness made him extra accountable for his actions. Because there would be no heirs to improve upon him or cover up for him, he felt he must "make the best of my existence so as not to put to shame the efforts of my forebearers." He was, incidently, extremely fond of children, in disciplined small doses, and had a knack for winning their affection. He never talked down to them and he amused them with his stories.

The ambiguities which afflicted Rock necessarily affected his relationships with others, particularly women. The dinner table bachelor with continental manners had few kind words for the opposite sex in general, regarding them with the mixed fear, suspicion, and contempt of a man of insecure sexual identity; at best he thought women strange and unpredictable; at worst, vessels of temptation. The women whom he befriended or admired were those who did not threaten him sexually, such as missionaries or the wives of friends who treated him with motherly or sisterly deference. Though there was nothing physically remarkable about him, women, with all the perverseness of their sex, were drawn by the very obstinacy of his solitude and his advertised unavailability. Rock's stories of China, marked by violence and acts of manliness, further increased his attractiveness in feminine eyes, but he tactfully kept all would-be seductresses at bay. Imagined insults to his sexual dignity, however innocently intended, raised his hackles.

Once, in Shanghai, in October, 1932, Edgar Snow entertained him at the Rose Room, a nightclub-restaurant known for its exotic dancers. "To me," the outraged guest recorded in an entry entirely out of proportion with the incident, "it was most disgusting, for everything revolved on a sexual pivot . . . so vulgar that the pen refuses to describe the scene"—but somehow accomplished the ineffable in graphic detail, breasts and pelvises included. "Had I known what Snow was taking us to, I would have refused. He had two American girls with us and, while he was dancing with one, I told the other to pack up and leave this town and go back to her mother." Two days later he was still berating Snow and the excesses of civilization upon which he blamed the odious displays of sensuality he had been forced, out of politeness, to witness. He protested the scene too much, more like a man aroused against his will than one in easy command of his passions.

Among the many thousand words that Rock penned in his diaries, those which expressed the extent of his loneliness are the most memorable; how much more absolute is the assertion that "I love nobody" than the whining complaint "nobody loves me." Rock's lack of love was not reserved for women. He meant exactly what he wrote, honoring the profound sense of the word love as though it were an emotion he thought he remembered but which had, inexplicably, evaporated from his life. The deprivation of loving pained him, and he searched for someone upon whom he could posit his affections, but his relations with people always defined themselves in other terms, and he found himself with acquaintances or charges, at best friends, rather than persons to love. Characteristically, he blamed others for falling short of his standards and failed to recognize that within him which resisted loving.

Rock romanticized the remnants of his family in Vienna to make up for his own lack of wife and children. Conveniently forgetting the violent quarrels of childhood, he idealized his sister Lina, wrote her frequently, and sent generous sums of money until financial setbacks made him nervous. Eventually he had the idea to take one of her three, now fatherless, sons under his wing, and he selected the youngest nephew, Robert Koc, for this honor. He arranged for Robert to meet him in Venice in December, 1933, during one of his absences from China and further proposed that, should everything proceed according to plan, he would take the boy back to Yunnan to help him with his work. Robert was a callow 19 and had never been out of Austria. He had not seen his illustrious uncle for twelve years and remembered him best for the presents he had brought. Bewildered by a language he could not understand and unaccustomed to luxury, Robert was dumbstruck by the Grande Hotel Danielli, where Rock had installed himself in splendor. Robert tended toward shyness under normal circumstances; Rock's dictatorial manners awed and embarrassed him; the peculiar mixture of generosity and meanness was confusing. Rock, who hoped to find a reflection of himself, was horribly let down. He decided Robert would never do for China and, on the pretext that the boy was useless as long as he did not speak English, returned him to Vienna. He would recall him a few years later when his longing to love was stronger than his memory of Robert.

Though Rock condemned Robert for weakness, it is hard to believe that he would have stood for anything but absolute obedience. Robert's older brother, Hans, a more single-minded individual, refused to take orders from his uncle. Rock, privately confessing to have more respect for Hans than for Robert, would have nothing to do with him. In fact, Rock could not bear to be contradicted by anyone and, therefore, surrounded himself with people who obeyed him, namely his Praetorian Guard—the Nakhi servants. For lack of anything better, they became his family in a medieval sense: he dispensed protection and justice; they, in return, complied with his every whim. Rock sometimes let this master-servant alliance pass for love, but he never succeeded in fooling himself completely.

There were people all over the world who liked him, wrote him letters, opened their homes to him, worried about his health, admired his work, loved his stories, enjoyed his company, and recognized his individual genius. He loved none of them, but their attention was vital to him. An energetic correspondent, he suffered when the Chinese mail service broke down and he was cut off from news from the West. Piles of Christmas cards arriving in China from many countries pleased him enormously, as though they had been valentines. But the level of intimacy is suggested by the fact that no one ever wrote "Dear Joe"; it was always "Dear Dr. Rock," "Dear Rock," or occasionally from Hawaii, "Dear Pohaku." In the society of Westerners—the American Consulate in Yunnanfu, a faculty home in Cambridge, Massachusetts, or a lavish patio in Honolulu—Rock found it easy to claim the center of the floor and, once there, was reluctant to leave it. Hostesses anxious to avoid conversational lulls could do no better than to invite him.

As time passed and he became a minor celebrity, Rock accumulated a long list of social acquaintances and colleagues whom he wrote or visited and of whom he thought as friends of one degree or another. Even those with whom he passed the most time and who, therefore, would have had the best opportunity to know him, describe him as a loner, admit to the feeling that they did not really know him, and recall that, with all his charm in company, he could be exceedingly difficult. They obviously put up with much more from him than he from them. Accustomed to ordering people about in China, Rock expected the same kind of service and attitudes in the West; his despotic reflexes troubled some Westerners.

Rock dreaded the loneliness of expeditions to wild places in China and, on a few occasions, invited 'another white man to travel with him with the idea of providing himself with like-minded company and English conversation. Without exception he regretted these decisions. In July, 1926, he engaged a young Pentecostal missionary, William Simpson, to accompany him to the Amne Machin. Simpson, an American, spoke fluent Tibetan and would have uses beyond simple companionship. It did not take long for Rock to find fault with him, however, and he complained that "missionary business cannot be carried out in conjunction with a scientific expedition, especially if that particular mission is of the hysterical kind. . . . While [Simpson] is kind and good-hearted in many ways, firmness is absolutely lacking in him. Its place is taken by too much brotherly love and sweet words while these ruffians here look upon such conduct as becoming to a silly woman." Simpson was not hysterical at all, but he was definitely critical of Rock's treatment of tribesmen and of his cynicism. Rock ordered him back to his efforts with the infidel and replaced him with a native interpreter. In 1929 Rock repeated his mistake with W. I. Hagen, a young man whom he lured from a post at the U.S. Consultate in Yunnanfu to join the National Geographic expedition to Minya Konka. Hagen lasted longer, but ultimately fared no better than Simpson. After trekking from Yunnanfu to Tali with Edgar Snow in 1931, Rock solemnly vowed never to travel again with a white man. He even declined one expedition because he had been proposed as a co-leader and knew the arrangement would never work. What started out as a pleasant camping trip with a couple of missionaries in a gorgeous meadow below the Likiang snow range finished by driving him nearly to distraction.

Rock liked to do things his own way and, as long as he had only to deal with Nakhis, Chinese coolies, soldiers, muletiers, he could demand, cajole, threaten and, if disobeyed, punish at will. White men, though always in the position of recognizing his authority over the caravan, often behaved against his wishes or argued with his decisions. Short of stranding them on the trail or sending them back, as he did Simpson, Rock was powerless to force them. In Simpson's case, Rock reasoned that he jeopardized the expedition; with Snow, whose sympathies for China's downtrodden were quickly developing, there were other problems. Snow started off on the wrong foot with him after their meeting in Yunnanfu, showing his ignorance by proposing to use potassium cyanide to disinfect the plants he would eat along the road "and kill himself in the bargain. *Sancta simplicitas!*" Nor was Rock pleased with Snow's frank criticism of the way he behaved with the Chinese. The journalist was embarrassed by the luxury of Rock's caravan, but admitted he felt safer in bandit-infested Yunnan with armed protection.[3] An eager novice, he submitted to Rock's style with good humor while annoying him with his youthful liberalism. Twenty-seven years later Snow wrote an amusing account of the march in *Journey to the Beginning;* Rock read it and enjoyed telling people that Snow had misquoted him. Rock summarized his feelings in one line: "never take anybody along but travel by yourself," adding a few nasty words for Snow. Yet, even after the Shanghai nightclub incident, he still considered Snow a friend. About the only Westerner in China who never provoked him was Pickwickian Dr. J. A. Watson, head of the China Inland Mission Hospital in Yunnanfu; Rock never traveled with him. The standard situation, however, was for him to detest the Westerners with him in the wilds and then, having dismissed them, to miss their company.

Rock's affections for his nomadic life-de-luxe fluctuated similarly, sometimes dangerously. That priceless ability not only to see the irony of one's situation but also to laugh at it was not often with Rock; more often he lapsed into morbid thoughts.

"If it were not a cowardly act," he despaired during the winter of 1926, "I would quietly shed this mortal body and thus once and for all end this hectic existence. Today for me was dreadful; it would not have taken much for my mind to give way. I am on the verge of a nervous breakdown. Not, perhaps, verge, but arrived at the collapse. . . . My brain is fatigued; I cannot concentrate. My mind worries over conditions in China, the impossibilities of getting any funds . . . All the roads out of Kansu are closed, mules are not to be had, and the fighting is coming closer. . . . Food is getting short, and nothing can be bought. Not a pound of flour is for sale . . . " Given this list of grievances, one can sympathize with Rock's despair. But suicide has its own terrifying logic, and lesser things drove him quickly to the end of his wits. A trivial disagreement with one of his Nakhis one morning made him write that he had "never been so voluntarily near death . . . " Such entries occurred with unsettling regularity in his diaries, and he himself feared that he might one day give in to the temptation to do away with himself. Accordingly, while living in Yunnanfu in the latter 1930s, he asked Paul Meyer, the American Consul, to keep his pistol for him as a precaution, explaining that he often thought about suicide. The Consul, rather amused by Rock's eccentric outbursts, agreed to keep the weapon.

It would be misleading to give the impression that Rock sustained a state of perpetual misery, because there were days when he was emphatically happy. He almost always awoke in high spirits in the morning. In the course of a few hours, his mood could undergo a total reversal as though, having experienced a transfusion, he had been drained of one emotion and replenished with another. As an example, the day after his quarrel with the Nakhi which brought him "so voluntarily near death," he recorded "all his happiness . . . " He experienced spells of well-being and wonder as he surveyed the awesome scenery in the undisturbed Chinese mountains. But, over all, the sense of loneliness and restlessness and despair, sometimes bordering on hyteria, prevailed. Others could translate their unhappiness to an intimate companion with a sullen silence or an angry glance; Rock had only his diary.

In further explanation of Rock's violent changes of mood it is also important to note that he was, for a good deal of the time, in indifferent health. His digestive system often malfunctioned, a condition no doubt aggravated by the polite necessity of showing one's pleasure by nibbling such dainties as hairy, rancid yak cheese profferred by the Muli king. Rock also had a weakness for rich desserts which overwork the sensitive stomach. In the 1930s the doctors diagnosed an intestinal block and, because he was inclined to be costive, prescribed daily enemas. In the 1940s he contracted an extremely painful facial neuralgia (*tic douleureux*) which nearly drove him berserk before he traveled all the way from China to Boston for an operation. So, though he lived to a ripe old age, Rock suffered physically most of his adult life, beginning with his early bouts with tuberculosis. Poor health tends to make people unpredictable and cranky.

The overwrought author of the diaries does not conform to the dauntless explorer who figured in Rock's social conversation. He gave people the impression that he loved China and his travels and, in fact, when he was dodging traffic at a New York intersection, inhaling noxious fumes in London, or submitting himself to the impudence of a waiter in a Paris bistro—in other words, when he felt victimized by what he called civilization—he did love the nomadic life. But when harassed by bandits or taunted by anti-foreign slogans in China, he swore at least a hundred times that if he ever got out of that horrid country, he would never return. At the moment of departure, however, he felt pangs of regret, a sensation common to other explorers. Reginald Farrar, noted that "the great drawback of all these expeditions is not the setting out upon them, but the deepening melancholy which attends their conclusion, as day by day you draw further away from the peace and lonely loveliness of the great ranges, back towards the frets and furies of that devastating and bankrupt futility to which we are pleased to give the name of civilization."[4] After a few weeks of Western civilization Rock ached to get back to the "peace" and "simplicity" of Yunnan, such inconveniences as brigands, soldiers, disease, and filth having temporarily receded to a corner of his mind. Back in China, he invariably castigated the country and its people and wondered what on earth had compelled him to return; he missed Western intellect and comfort. The pattern repeated itself so many times that it would be ludicrous if one did not detect in it a familiar anguish, a relentless search for serenity. While ordinary people do their existential suffering on a treadmill between office and home, Rock did his between China and the West. At least he recognized his own masochism, what he called "a desire to suffer mental agony" thereby augmenting it, and perhaps his grasp of this irony saved him.

ROCK'S ATTITUDE TOWARD THE CHINESE

Rock's life in China was resolutely Western, a style much more difficult to achieve in the Tibetan borderland than in the foreign compounds of Canton or Shanghai. In this respect he resembled the majority of Western residents but, unlike them, he was well versed in Chinese history and culture and, when the situation demanded, observed local customs with a tact and precision that few Westerners attempted, much less attained. What is curious, by contrast, is that instead of producing the sympathy and understanding which came to most foreigners who made the effort to learn about China, Rock's historical investigations had quite the opposite effect. As far as he was concerned, the greatness and glory of the Middle Kingdom had been permanently extinguished by the Republic. Though he ingested volumes and traveled widely, he failed to achieve the historical perspective or compassion that one would imagine in someone so familiar with a civilization. Herrymon Maurer, an American who spent the early years of the Sino-Japanese war teaching at West China Union University in Chengtu, furnishes an interesting contrast of attitudes in his book *The End Is Not Yet*. In the chapter called "Coolie Democracy," Maurer interpreted the life of the Szechwanese chair-bearer in romantic terms; but where he saw great human dignity, Rock saw meanness, filth, dishonesty, and indifference.

The longer he stayed in China, the more Rock grew to despise the Chinese. The scenes of brutality and human degradation which he witnessed so frequently reinforced his negative opinions. Rock attributed his loathing for the Chinese to his love for the Nakhi and other mountain tribesmen whom the Chinese had oppressed for centuries. But despite his repeated admiration for what he misconstrued as an idyllic existence, he thought of the tribesmen as children or graceful animals and treated them accordingly; and for all his ranting and raving about Western civilization, Rock remained absolutely convinced of the white man's superiority.

Rock's racial bias, often typical among European and American residents of China, was of the same vintage as that of the archetypical British colonials in India or the French in Indochina; it was the same kind of prejudice that inspired a colorful vocabulary among American soldiers during World War II and added such words as geek, chink, and slopehead to our lexicon of derogatory names for ethnic groups; still later, it was the same attitude that permitted the Pentagon to announce American casualty rates independently of those of our South Vietnamese allies, as though the latter pertained to a different species. And, in the interest of fairness, it was racial bias that made the Chinese call Westerners long noses and red-haired barbarians, or the Tibetans equally unflattering names, the distinction being that prejudice on the side of power is always more brutalizing. Racial prejudice is, however, an ugly fact of life. It comes as a surprise in Rock only because one expects it to be tempered by experience and intelligence. No doubt the expectation is unreasonable.

Rock treated the Chinese and tribespeople according to his opinion of them. Samples of his kindness abound. When he left Choni he distributed money to children and beggars, after screening the opium smokers; he bargained with the Muli *tussu* for the release of prisoners from a hellish underground dungeon; he ministered to the sick as skillfully as most medical missionaries and asked nothing but safe passage in return. But he also bullied the people of China and was not above striking a blow or two with his riding crop to rouse a lazy muletier. "One of my Szechwan chair coolies got fresh," he remarked one night, "and I promptly showed him where he came off. He cringed and cowed like all Chinese cowards do when confronted by anyone who is in earnest . . . I will stand for no impertinence, no nonsense from anyone in this country." In the name of dignity and face, he strutted before his hosts, and his manner, coupled with what looked to the Chinese like fabulous wealth, won him, according to Edgar Snow in *Journey to the Beginning,* the epithet of "foreign prince," among the Yunnanese, a step up from the "count" of his adolescence. After he had traversed some new terrain, he gloried in having been the first white man to do this or that which thousands of Tibetan nomads had done for centuries. It is not intended to downgrade Rock's extraordinary feats but, rather, to observe that his "first white man" boast implied that anything which had not been accomplished by white men simply had not been accomplished at all.

Rock's assumption of superiority as a Caucasian supplied him with a power in China which it was impossible for him to duplicate in the Western world. And it was as much the lure of this power as the rich botanical fields or cultural novelties that acted like a magnet on him. Where else but in some alien land, peopled by a "lesser" breed of man, could Joseph Rock, motherless son of an ill-tempered Viennese household menial, play the whimsical potentate? In China he received daily reassurance of his authority as his Nakhis scurried to do his bidding, coolies cringed at his curses, and district magistrates and tribal kings offered their best, however repulsive, in his honor. So, too, his acts of generosity and pity further proved his power to bestow small favors upon the needy. He made these gestures regally, never with humility. One begins to understand, then, his inevitable displeasure with white traveling companions. They represented threats to the power which he held so precious and which, in the recesses of his mind, he doubted. No Chinese or tribesman dared contest his will; for a white man to challenge him in their presence was as much as a physical assault. To keep his fragile authority intact, Rock had no alternative but to live and travel in exile.

The same tactics which made him a prince in China got him nowhere in "civilization." His money, a princely sum in Yunnan, became painfully ordinary on the outside. He could order custom-made shoes, silk shirts, and tailored suits to dress in conformance with his idea of himself, or stay at the best hotels—

which he invariably did, but he did not have the satisfaction of discharging orders to servants or armies of escorts. People looked askance when he returned complete dinners to a restaurant kitchen on the pretext of some trivial imperfection and then, to top it off, complained loudly to the head waiter. His efforts to substitute the servility of the Nakhis with that of his nephew failed miserably. With his sister Lina and her sons, who were beholden to him for monies delivered, he could most nearly reenact the role of royal benefactor which he played in China and reaffirm his self-image. But, in "civilization," Rock ultimately felt himself impotent. Even Lina would quarrel and disobey orders; she insisted, for example, on selling the country house he had purchased for her and returning to cramped quarters in Vienna. In his travels in the West Rock hobnobbed with important, wealthy people who sat spell-bound by his tales. They, however, were usually more powerful in a Western context than he, and he soon became restless. He returned, for as long as it remained possible, to China, where he had less competition choosing, unlike Cato, to be first in Utica rather than second in Rome. The price he paid for his choice was solitude.

Rock's was the loneliness not of geographic position but of power. Peter Goullart's reminiscences of Likiang are filled with people: family situations in which he became involved, young workers for whom he cared deeply, and a sense of identification born of working with the Nakhis to help them solve their economic difficulties. Rock, in the same setting, responded differently, confronting every new situation with the will to dominate, constantly testing his power lest he should wake up one morning and find it gone. He conceived a strange and bitter definition of personality. "Personality cannot be developed in solitude," he observed. "It must be surrounded by other living beings to make itself felt. What we often call personality is the power to perceive the weakness in others in a twinkling of an eye and thus dominate by sheer arrogance and self-forwardness like a rooster charging a council of hens among a few grains of wheat, scattering them, and then picking up leisurely the rest. The hens look on with cowed heads."

However small the domain, the man who rules it is classically lonely. He cannot afford the luxury of loving nor can he bare himself to love. He must be served and see evidence of service. The uninterrupted assertion of will comes easily to few and only to those who have supreme confidence in their power. The man who lacks that confidence is vulnerable. Thus Rock could be thrown into a depression by something as small as being seated "below the salt" at a dinner party or a falling out with one of the Nakhis. He believed he could trust no one. Under the strain of his pose he thought of suicide, a solution which, according to his religious training, was untenable and cowardly. The alternative was total withdrawal from society, a commitment to a meditative life close in form if not in content, to the priesthood he had rejected as an adolescent. The notion of monkish asceticism appealed to him because it held the potential for the serenity which eluded him. But his attachment to worldly things, his restless curiosity, and his enormous pride came into conflict with permanent retreat and, while he remained sexually abstinent and in other respects self-disciplined, he never got beyond meditating about meditation and feeling somewhat uneasy about being a man too much of this world.

Every thirty-first of December Rock renewed his covenant with God in a solitary ritual. As midnight approached, he fell to his knees and prayed aloud, "*Thy* will be done, *not mine*," copying the words in his diary and then signing his entry as though it constituted a legal document. Having been turned back at the Vatican for want of a proper suit many years ago, he had long ceased to think of himself as an adherent to any organized religion, but he claimed to practice Christian ethics. Except for New Year's Eve, the name of God did not form easily upon his lips; the god he dealt with as an adult was the Great First Cause that he perceived in nature. Only in confronting nature at its most dynamic did Rock submit to a power greater than himself. These were the precious moments when, for once, it was he who was passive, when he was freed from the effort of making himself felt by others. Only a few days after he had been ready to commit suicide in Choni he chanced on one of those vistas that moved and restored him. "It was a pleasure to be alive and to breathe in this grand mountain air and to let this bold, magnificent scenery react on one's soul which, in my case, said how foolish it would have been to end when there is so much to enjoy and to behold in the grand temple of the First Great Cause. One's heart rejoiced to be permitted to view such mighty works of nature, especially as no other foreigner had thus far been privileged to visit this hallowed shrine."—again the 'first white man' refrain. The Victorian prose that Rock affected to celebrate his god's work expressed an appreciation in the German cultural tradition. This is not the fragile nature of Wordsworth, the bittersweet nature of Keats, or the stylized, delicate nature of Chinese scrolls; it is nature in macrocosm, bold and exaggerated in outline, grand and sonorous like the music of Wagner and Beethoven or the poetry of Goethe. Nature's subtleties, the intricacies of its art, did not move Rock. What he loved in nature was power—thus the use of grand, bold, magnificent, and mighty. His idea of god was his extended self, power perfected.

Sometimes god, nature, life, and death all seemed to make sense to him, but everything came out garbled. One morning in 1931, for example, he recorded: "Everything was so lucid and clear in my mind and I felt as if I really comprehended nature, its laws and mysteries. Alas, such lucid moments do not come often . . . [and are like] flashes of light which pass in a moment and leave one in deeper darkness . . . It is the expression of the conflict of mind and matter—the former infinite, all comprehending; the latter limited. It is the separation of mind from matter which we call death which may be, after all, entering into the portals of knowledge or, better said, the realization of all knowledge, yea Knowledge itself. . . . Existence means limitation; death, freedom from existence and hence, perhaps, omniscience." And so, death tempted him once again, promising what life did not deliver. The threats of the fire and ice of hell which he had heard in his youth did not terrorize him, nor did he really believe in the wandering soul he had proposed as a solution for his lack of progeny. He looked forward to death to take him from life which was, for him, more pain than pleasure. Until then he submitted annually to god's will under which pretext he could exercise his own and lived out his destiny in seclusion, wandering like Nietzsche's Zarathustra:

"Now as Zarathustra was climbing the mountain he thought how often since youth he had wandered alone and how many mountains and ridges and peaks he had already climbed.

"I am a wanderer and a mountain climber, he said to his heart; I do not like the plains, and it seems I cannot sit still for long. And whatever may yet come to me as destiny and experience will include some wandering and mountain climbing: in the end, one experiences only oneself. The time is gone when mere accidents could still happen to me; and what could still come to me now that was not mine already? What returns, what finally comes home to me, is my own self and what of myself has long been in strange lands and scattered about among all things and accidents."

VI

SEEKING THE MOUNTAINS
OF MYSTERY

"Looking for trees or even shrubs on the Kokonor [plateau] is like looking for a needle in a haystack," Rock reported gloomily to Professor Sargent in October, 1925. Two weeks later, having poked about the eastern reaches of the Richthofen range, his description to Sargent was even gloomier. "I have suffered a good deal from the cold of these high altitudes and from the severe, cold northwest winds. I also had an accident with my left hand, which has been badly swollen for several days and has been quite painful; . . . It will not be possible to go the whole northern length, or rather to the extreme west of the Richthofen range on the northern slopes this year as it is getting too cold, and camping at such altitudes in such a latitude, exposed to the cold winds of the Mongolian desert, is more than I can at the present endure. Had political circumstances permitted my coming here earlier, all could have easily been done, but the fates are against us. This region is terribly expensive. Grass or hay is sold by the pound. It is also difficult to hire animals to go to these out-of-the-way places, as the muletiers are used to travel only on the big roads from inn to inn. It is hard on the animals to spend night on night out in the open snow and ice with temperatures as low as 2° and 4° Fahr. and with inadequate fodder, as often as not one is forced to camp where there is no grass. Camels cannot go where there are ice and steep hillsides, as they slip easily and break their legs. Quite a problem, and the only way I could get a muleman to go was to commandeer them through the officials and state we are going there and there, and finish; no word can be said and they must go. They whine continuously and groan so that it gets on one's nerves; and to make things unpleasant they go so slow that one is glad to get rid of them when possible. They are all alike."

When Rock reached Kanchow (Changyeh), a town near the Mongolian border, he also had reached the limits of his optimism. This whole venture northward, undertaken at an enormous outlay of energy and finances, was proving botanically fruitless. He had made a respectable collection of birds near Lake Kokonor, but in all other respects was disgusted with the results. After days of trekking in frozen monotony north of the mountains, he had looked ahead to Kanchow for visual and mental relief; instead he found a wretched hell-hole where, due to the necessity of exchanging his mules for camels, he was obliged to stop for a week. He was not in a mood to see things with friendly eyes. What he observed he committed to his diary.

"The town," he reported, "is sorely disappointing—dirty streets full of dust; people, especially children, wallowing in dust. They are absolutely indifferent to the dust. They seem to take pleasure, judging from the way they walk, [in stirring] up clouds of the microbe-charged dust. People stare at one and do not know what to make of it all. Every house and wall is cracked by the earthquake. Ponds with tall weeds take up a good part of the town. Not ponds, really, but swamps. Mosquitoes are said to be fierce in the summer.

"The Moslems hold sway here, Ma Lin being the defense commissioner but now absent. The soldiers, all Mohammedans, oppress the Chinese terribly. At every gate are Mohammedan soldiers, who rob the peasants as they enter the city. From one they take a sheep, from another a horse, and if [the peasants] remonstrate they are beaten and kicked; they are tied up and pulled up backwards, hanging from posts or ceilings. The soldiers here are absolute robbers. This admitted even my Mohammedan Sining escort. The magistrate, whose name is Yang Yen-ch'uan, I as much disliked as he holds with the Moslems. He is not friendly and, although he called first, . . . probably out of curiosity, he showed himself not very friendly; and when I returned the call sent word he was not at home, although I knew he was . . .

"Indifference, the great sin of the Chinese, stares at one from no matter where one looks. Cowardice is the very incentive to the Moslem boldness. They put up with everything. Children will sit in carts behind the mules waiting for the manure to drop . . . There is lots of competition even in that line. Little can be bought in the town, and what there is is frightfully expensive. . . . Chinese do not know comfort. They run about with overcoats in the house; no fire and the temperature freezing. All water has to be brought from some common pool where pigs, children, dogs, cows, and horses drink . . . or from some common well where everyone dips his own bucket after putting it on the dirty sidewalk or street. . . .

"Kanchow is one huge crows' nest. In the evening millions of crows come flying from the plain over the old Kanchow wall cracked by the earthquake to roost on proper trees."

By substituting mud for dust and adding an assortment of stenches, subdued by Kanchow cold, one has Rock's view of the average western Chinese town, a place with virtually no redeeming qualities. From Mongolia to Burma the differences were academic: politics, climate, building materials, and architecture might vary, but filth was the ubiquitous common denominator, afflicting Shan, Chinese, Tibetan, Moslem, and Mongol without distinction and offending Rock's fastidious senses. For every town he praised, he passed another twenty that he deemed unfit for civilized humans. Kanchow was just one more like the rest. Rock resented wasting the week there, rummaging through his plant and bird specimens to blot out the unpleasantness of his surroundings. He grew morose and longed, once more, for death which, predictably, did not oblige him. Instead he continued his aborted venture.

After leaving the disappointing mountains, Rock turned back toward Sining on the last day of October and crossed the Kanchow plain. The Chinese and Mongols who worked the poorly irrigated land, unaccustomed to such imposing caravans, shrank from the party, taking them for Moslems or brigands on the rampage. Rock observed their terror callously. Winter closed in fast now, and he was desperate to get to Sining and on to Choni where he would stay for the winter, before the snow became deep enough to block the mountain passes in southwest Kansu. Already the cold was unbearable, and he had been through one blizzard. He had chillblains in his injured hand, and his ears were frostbitten.

He found Sining in utter confusion. He had expected preferential treatment from Ma Chi-fu, but the Moslem chief was incommunicado, pondering ways to counter a new Chinese regime in Lanchow, which had taken an unwelcome interest in his activities. Hostilities between Moslems and Tibetans were still feverish, the former having ransacked the lamasery at Heitso during Rock's absence in the mountains to punish the Tibetans' earlier burning of Moslem houses. The Sining military establishment, fearing a showdown against an alliance of Tibetans and Chinese, was seizing every pack animal that wandered into the district. Rock's request for mules fell upon disinterested ears until he produced some cash. Soldiers could not be spared for escort, he was informed; but he was, by now, an accomplished bluffer. He threatened the Sining officials with advising the American representative in Peking that he had been refused protection, and left the city at dawn the next morning with only his Nakhis and pack animals. It was only a matter of hours before Moslem soldiers caught up with him on the trail, demonstrating that extrality was still taken seriously in some places.

In the wilderness again; "Mountains like the skeletons of huge dinosaurs or dragons," Rock wrote in November, 1925. He forgot the cold for a few hours and stopped in Labrang, which the Moslems still controlled. Simpson, the young Pentecostal missionary, filled him in on the gory events of the last several weeks. A few days later Rock passed through Heitso and confirmed the accuracy of the report; the lamasery had been picked clean. By December 3, he reached Choni. Prince Yang, alerted to his arrival by an advance man, rode out beyond the gate to greet him. The wanderer had, so-to-speak, come home.

Rock settled into his quarters, rested, and worked over his collections, reflecting that, because of the good botanizing in Tebbu country, the year as a whole need not be dismissed as a fiasco. On Christmas Eve he retired early and privately reminisced about Christmas in Vienna, but bitter memories crowded his mind and he fastened on his mother's death instead of the dazzling Christmas tree. He celebrated New Year's Eve by inviting Derk and Hulton, the missionaries for whom his diaries usually had biting words, to share in a feast of spaghetti with cheese, roast loin of pork, applesauce, roast potatoes, sourkraut, green peas, mushrooms, asparagus salad, cookies, coffee, and creme de menthe. He then dismissed his guests before midnight so as to renew his annual pact with God in private. "Thy will be done, *not* mine," he wrote with customary flourishes meaning, in particular, his trip to the Amne Machin, about which he was pessimistic. Chinese New Year came and went with few formalities. Rock's Nakhis lined up in his quarters early in the morning on the first day of the new year, bowed low, and wished him well; he bestowed small gifts of money upon them.

Rock hibernated in Choni until the end of April. Except for excursions to Taochow for supplies and mail, and a wasted attempt to check conifers in Tebbu country, he occupied himself with mailing specimens back to Boston and observing the more curious practices of the lamasery. Choni's great annual event was the butter festival, a religious celebration for which the lamas fashioned huge, elaborate images of their idols out of brightly tinted yak butter. Butter for the idols was collected during the autumn, and the lama artists began their work in mid-December, taking competitive pride in their sculptures. The preparations entertained Rock, who passed the time taking notes on the ceremonies and watching the lamas at work. The *National Geographic Magazine* published his article on the Choni butter festival along with several photographs, including sixteen in color. The next year the *Illustrated London News* picked up the story under the title "Butter as a Medium of Religious Art."

Rock spent long hours with Prince Yang learning about the little principality, but saw less of the two Western missionaries than one would have supposed. Simpson descended from Labrang occasionally, Cassandra-like, bearing the latest bad news. By February Rock knew no more about the trip to the Amne Machin than he had in December.

THE POSSESSED DANCERS OF HOCHIASSU

A single unearthly experience marked that winter in Choni. One day in February, Rock, two of the Nakhis, and Prince Yang went to witness dances in the neighboring lamasery of Hochiassu. Rock had heard rumors from the Choni villagers that the Hochiassu ceremonies were something special, and, with the idea of photographing them, persuaded Yang to accompany him as an interpreter. The Prince, who had never been to the ceremony, agreed reluctantly. Rock, seated beside him, settled down to his note-taking when strange things began to happen. As he recorded the experience in his diary:

"I looked up from my notebook and, looking down upon the dancers in the crowd, I saw three women on the ground, kicking and fighting with four men who were holding each woman. They cried and shrieked and tried to get loose and rush at the dancers. People began to laugh at them and while they were laughing they, as suddenly as can be conceived, were taken by that unknown something and were in the same condition as those they had laughed at before. I felt most peculiar and said to the Prince, 'I will go down and take some photographs of the dancers.' When I arrived in the courtyard I wandered aimlessly from one place to another and finally returned to the chanting hall, where I opened my camera . . . I felt much better there but soon felt uncomfortable."

"As I was looking around I saw the Choni Prince come down pale as death. He saw a lama bring up a tray with tea and was almost about to pass the tea [when] he was seized by that indescribable something and began to shake from head to foot and had to be taken charge of by the other lamas. The Choni Prince thought it was time to go. He felt terrible. . . . I began to feel a bit uneasy, but as I was about to take a photo the skeleton dancers had disappeared . . . Mr. McGilvrey [a missionary] who was standing in back of me felt terribly oppressed. One of my boys, Ho Chi, touched him on the shoulders and said 'Please tell Mr. Rock that I want to go back to Choni. I feel terribly bad and cold. This is a very queer place!' I said, 'Yes, go quickly.'

"At that moment I felt a most peculiar sensation, as if I was giving way to some unknown force, and I felt myself become powerless. I managed to say 'I must go, I cannot stand this.' The dancer whirled around furiously, the possessed squirming on the ground, crying, screaming, weeping, and tearing their garments. The men holding them were almost powerless. All in all twenty became possessed. I left the chanting hall and its courtyard as quickly as I could, but lamas insisted that I go to the Tsung-kuan's [chief lama] house and partake of refreshment. The Prince followed and was exceedingly glad that I did not care to stay any longer. I walked into the small reception room, but I had hardly sat down when the powerless feeling came over me in a few minutes, and my sight was practically gone. I could ill distinguish things. I tried to repress this feeling and started deep breathing, but every second I felt myself going and I came to the conclusion that if I did not leave immediately, I would be unconscious the next minute. So I crawled down the carpet past the Prince who was only too glad to have an excuse to leave . . .

"As soon as I got out of the monastery compound I felt relieved at once. I confess that had I stayed either in the courtyard where the dancing was going on, or in the room where we were to have the feast, I would have succumbed to this outside influence, whatever it was; of that I am certain. I may say that never in all my life have I felt in such a sudden manner my will power and my control over my being leave me . . . It was like a filtering of something through one's body which took charge of one completely . . . It was the elimination of self and the control of self through another force, which I fought, but to which I would have succumbed . . . The lamas explain this as the god of the monastery taking possession of one's body."[1]

On the way back to Choni, a deep serenity replaced Rock's malaise. Yet several hours later when he was copying descriptions of the dances into his notebooks, he felt queasy again, started to shiver, and had to stop writing. He had gone to Hochiassu a skeptic, smug in the assurance of his self-control and believing that only the superstitious people of the mountains could be prey to "possession." He came away more fascinated than upset by his own susceptibility, and no doubt McGilvrey, the missionary, was more at odds with himself than Rock who, as a good scientist, cast about for rational explanations. He had eaten nothing except watermelon and raisins. He did not mention the possibility of intoxicating fumes but, with his sensitivty to odors, he would have been certain to notice anything peculiar. "There is a power or a spell of some kind," he concluded unscientifically to Ralph Graves, and thereby ended his diagnosis.[2]

By the time of the butter festival, a few days later, Rock had ceased pondering the weird happenings at Hochiassu. He photographed the ornate butter gods and studied the ritual dances which, unlike the dances at Hochiassu, resembled theatrical performances and produced no writhing audiences. Ceremonial diversions notwithstanding, Rock grew restless biding his time in the full little town of a few hundred families, and inactivity made him impatient. He flew into a rage upon discovering one of the Nakhis with opium, immediately ordered the offender back to Likiang, and with his riding crop beat the Choni inkeeper who had sold the drug. He circulated word that he would punish anyone who sold opium to any of his men.

The unpleasantness made Rock more anxious than ever to resume his explorations, but now, in addition to the provincial political problems which had thwarted him for months, he found himself dangerously low on cash. The expedition had been costly from the beginning due to inflation and the necessity of hiring huge escorts. In November he had appealed to Sargent for an extra $2,000 and soon after had asked for an advance on his salary. By February he had received neither a check nor an acknowledgment of his requests and he was desperate. As though to give evidence of his needs, he offered Sargent his picture of the Kansu currency imbroglio:

"Every town in Kansu has its own scale for weighing silver and, in addition, they have two kinds of weights, the old and the new weight; the latter are naturally lighter than the old weights. There are in circulation Yuan Shih-k'ai dollars of the 3rd, 7th, 8th and 9th and 10th year. The third year dollar is the current dollar taken at par. The others of the 7th to the 10th year are all discounted for two reasons: first because Yuan Shih-k'ai died in the 3rd year of his rule, hence there cannot be a 10th year Yuan

Shih-k'ai; this is a superstitious reason. And, second, all the issues . . . after the 3rd year are a little lighter in weight. If the 3rd year dollar has a scratch or a mark on it it is discounted five cents. Then come the Manchu and Peyang dollars and the Dragon dollar of the different provinces, each of which has a value of its own. The present day Szechwan dollars are not accepted in Kansu. In addition there is the Lanchow paper dollar which has its own fluctuating value. So if one buys anything one is asked with what kind of money or dollars are you going to pay and the price is then fixed accordingly, the apparent standard being the 3rd year dollar Yuan Shih-k'ai.

"Now these various dollars are exchangeable into coppers only. There are no small silver coins in Kansu. Kansu has no copper so all the coppers are imported from Szechwan. The road being a long one makes the coppers expensive. In Chengtu a dollar exchanges for 4,000 cash, but that fluctuates. In Choni the dollar exchanges for 2,300 or 2,500 cash. No larger copper pieces are taken than 20 cash pieces. Now this [means] however, only yellow brass coppers, so to say. If you want red coppers you get only 1,700 cash. This all does not mean that you actually get 1,700 in 10 and 20 red coppers, or 2,300 cash in 10 and 20 yellow coppers, for the Chinese reckon 85 to 95 actual cash for 100 so-called cash. In Labrang 100 cash means really only 85 cash, and in every village the dollar is exchanged to different cash values. The Tibetans out here are still so antiquated that they refuse dollars but will accept only sycee or lump silver. This means you have to run around with *a scale*, hunt the scale in use in that particular locality and an axe to chop the silver. In paying with silver one always loses as one has to take the other man's scales, and where he puts the weights it is heavier than where he puts the silver. This is a recognized custom . . .

"The currency system is enough to drive one mad," Rock wrote Sargent on Feb. 12, 1926. "I think by the time I get through here I shall be fit for the post of Minister of Finance." Sargent, a Yankee of the string-saving variety who kept his millions in impeccable order, must have been struck dumb by this bewildering bookkeeping. He had, incidentally, wired two checks but, out of a well-meaning desire to get them to their destination as quickly as possible, had sent them via a bank other than the one instructed by Rock and thereby, inadvertantly, added another knot to the tangle. Rock, meanwhile, victim of the sluggish postal system, knew nothing of the error and waited suspensefully. Hoping that money would arrive, he now made regular trips to Lanchow where he befriended the Postal Commissioner, who regularly informed him that his funds had not yet appeared and gossiped knowledgeably about provincial politics.

The warring factions of Kansu had arrived, temporarily, at a stalemate. In March, 1926, General Ma still held Labrang, and the Living Buddha thereof, spitefully refusing to bestow his holy presence on the lamasery, continued to hide among the nomadic tribes in the mountains. The Tibetans, loosely allied from the start, had disbanded to lick their wounds after the humiliating sack of Heitso; but, if anything, their hatred of Ma's Moslem troops and machine guns had doubled. They awaited a propitious moment to regroup. General Ma did not fear the Tibetans. What worried him—and had been the source of all the confusion during Rock's last trip through Sining, were rumblings from the Chinese.

RIVAL FORCES IN KANSU

An entirely new presence had assumed the upper hand in Kansu. The same Feng Yu-hsiang who had charge of Peking when Rock returned to China in the fall of 1925, had been forced out of the capital by a coalition of rival warlords, Wu P'ei-fu of Chihli and Chang Tso-lin of Manchuria. He had retreated northwest with his Kuominchun, or National Peoples Army, first to Kalgan, and by early 1926, had divided his forces between Shensi province and Lanchow in Kansu.[3] The introduction of the Kuominchun into Kansu struck fear into the heart of Ma Chi-fu and significantly altered the tensions in the province. Feng, in his flight, had brought all his baggage, including the politics and problems of eastern China, and thus ended Kansu's provincial innocence. Before his army arrived, the hostilities, however bloody, had been quaintly local; Ma Chi-fu and the Living Buddha of Labrang had engaged in their private quarrel, oblivious to treaty ports and Peking. Now, for the first time since he had entered Kansu, Rock saw signs of anti-foreignism, an import from the coast. White Russians who had escaped to China and were not protected by extrality got the worst of it; in Lanchow they were thrown into jail, and some—former military officers—were shot. Rock, always ill-disposed toward Chinese generals, scoffed at Feng's "Christianity," labeled him a communist, which, in fact, he was, and wondered what the new regime meant to his expedition.

In some respects Feng Yu-hsiang was early vintage Mao: a moralizing, genial, hard-working, strapping man of six feet, he had advanced from army private—a rank the Chinese compared unfavorably with scum—espoused Christianity with Cromwellian fervor and, in the early 1920s gained the affections of foreign missionaries and the attention of the press, which played him up as a peasant hero. He ran a tight army, enforcing Puritan morals, though not so stringently as to eliminate opium smoking among his troops. His image found particular favor with Americans who, for a long time, viewed him as the most agreeable contender for power in China. Feng's humble origins and military excellence, if not his religious eccentricities, also endeared him to the Soviet Union which, while having relinquished extraterritorial rights, by no means had abandoned its designs on China. There were many Russian fingers in the Chinese pie, and Feng was, one of them. In contrast with some of his contemporaries, Feng's conversion to communism was opportunistic rather than intellectual. Eager for the aid the Russians tendered, he accepted the ideological strings attached. By 1925 it was no secret that he was receiving extensive Soviet assistance, and many of his American fans became disenchanted. American thinking about communism had not yet petrified into obdurate opposition, but Feng's activities, which included revolutionary indoctrination in Moscow in 1926, created anxiety in diplomatic circles. When his 200,000 or more troops withdrew to the northwest, United States officials observed with an indefinite mixture of relief and apprehension.

Feng had selected Lanchow as the ultimate western retreat because the Russians could funnel arms across the Mongolian border with practically no interference; meanwhile he kept the eastern warlords at a comfortable range. When they learned that the National Peoples Army was headed in their direction, both Tibetans and Moslems assumed he had been sent by Peking to punish Ma. This was wishful thinking on behalf of the former, downright fear on the part of the latter, and an indication of how little either party understood east coast politics.

Feng's instrument in Lanchow—the Christian General himself being otherwise occupied for the time being in Kalgan—was Liu Yu-fen, who had arrived in October, 1925, with Kuominchun soldiers. With Feng's approval and with elaborate guile, he arranged for the capture and assassination of ailing Governor Lu's chief subordinates (including Li Ch'ang-ch'ing, whom Rock had so disliked) at a banquet and, having gotten a foothold in the capital, proceeded to wheel and deal for the rest of the province, as related by Sheridan.[4] Ultimately, of course, Feng had more important things on his mind than control over inhospitable mountains and steppes; he pointed his nose eastward, whence he came, and more or less ignored the Moslem stronghold in Sining. In deference to his backers, he played the communist revolutionary, consolidated his forces, and waited for an avenue back to power.

Feng's army was fatigued. "The Feng Yu-hsiang regime is in a bad way," Rock wrote Sargent in April, 1926, "and they are attacked all aound. The Governor of Lanchow has asked General Ma of Sining for 30,000 soldiers, which he is not sending. The various factions are just awaiting the tide of events in the east and then will pounce on Lanchow and drive the present regime out. As the various factions contending for the military governorship of Kansu are by no means in accord, there will be fighting when the present regime will have been thrown out, and perhaps a Chinese-Mohammedan war will involve this province." Meanwhile, since they misinterpreted his motives, Feng's troops made everyone in Kansu uneasy. The province was taxed unmercifully to support the army. Ma Chi-fu made war preparations; the Tibetans waited impatiently. A Chinese warlord in Pingliang, in the eastern part of the province, thinking that Feng would challenge him, rattled his sabers provocatively. And, as news spread of the mistreatment of the White Russians in Lanchow, the foreigners felt insecure.

Rock decided to cast his lot with the Tibetans on the theory that if any of the possible battles erupted he could, like the Living Buddha of Labrang, hide out among the nomad tribes in the mountains. He languished irritably in Choni to see what, if anything, would happen and to wait for notification that his funds had finally arrived from Boston. Nothing materialized, either politically or financially. He received a three-month-old letter from Sargent on the first of April, but it contained no news about his checks. Early spring plants bloomed around Choni, forcing him into a decision. In his restlessness, all the conditions which had constituted obstacles the year before—Tibetans vs. Moslems, bandit, difficulties of securing pack animals—not to mention recent developments, seemed less formidable to him. After months and months of hedging, he suddenly announced his intention to go to the Amne Machin and the gorges of the Yellow River.

EXPEDITION TO AMNE MACHIN

Rock temporarily solved his fiscal quandary by writing a check for a few thousand dollars on his personal account and putting it up as security against a loan from Prince Yang, a man evidently sufficiently sophisticated to negotiate in symbols. Though Ma Chi-fu had originally promised him safe conduct in the mountains, Rock gambled on the Tibetans and kept his distance from Moslems. Since the advent of the Kuominchun, he could not be sure that Ma would honor his offer and he did not want to travel all the way to Sining for nothing. Furthermore, for a good part of his trip he would be in territory controled by the wild and wooly Ngolok tribe whose hatred of Ma Chi's men would probably make a Moslem escort more of a liability than an asset. Bandits were Rock's chief worry; "One must take . . . for granted that every Tibetan, at least in this part of the world, was a robber sometimes in his life," he wrote Sargent, and that even lamas are not averse to cutting one's throat although they would be horrified at the killing of a dog or, perhaps, even vermin." He had no desire to introduce any extraneous perils. He arranged, therefore, for twenty mounted, armed Tibetans to take him as far as Ragya; there he would improvise for the next stage. The letters he had secured in 1925 from the Living Buddha of Labrang would presumably give him bargaining power.

He mounted and armed his Nakhis, packed up thirty-three

mules that had been troublesome to procure, and proceeded to Labrang on April 29. There he engaged sixty yaks—better for the terrain, though a yak could carry only two-thirds of a mule-load—and stocked up on provisions for seven months: 2,000 catties of flour, salt, native soda for baking bread, beans and peas for the animals because the grass was not up yet in the frozen country to the west, and his more luxurious personal items. "West of Labrang is nomad country," he explained to Sargent, "and the only thing which might be bought is yak butter which grows whiskers, sour milk, and mutton." He also bought bolts of raw silk, cotton, and satin, thread and needles for bartering purposes; satin brocade as gifts for the Living Buddha of Ragya and tribal chiefs; hundreds of photographs of the Dalai and Panchen lamas and *kataks* (salutation scarves made of silk; the longest and best are always presented to the most important persons) as presents for lamas and tribesmen, the photographs being especially treasured items.

The necessity for elaborate preparations and his own wariness of the trip caused Rock to comment uncharitably about an account he had read in an old *National Geographic* of Wilson Popenoe's expedition in South America. "It sounds to me more like a picnic than an exploration. To make a trip from the end of the railroad with a knapsack on his back to an Andean hostelry, an afternoon's journey, and walk into the woods along the highway is hardly worthy of the term 'arduous journey.' . . . Out here the traveling is just a bit different."

On May 4, the night before he was to leave for Ragya, Rock felt depressed. He was sick of buddhas, prayer wheels, and lamaseries and longed for a comfortable American home.[5] He was not afraid, only numb, lonely, alienated, and poised to move further into isolation rather than out of it. Perhaps hoping to relieve his loneliness, Rock started from Labrang with the earnest William Simpson as translator and traveling companion, but the latter's good heart soon proved more of a burden than his familiarity with Tibetan dialects did a blessing. He made himself constantly likeable and constantly troublesome, pitting his gentle democratic ideals against Rock's steely authoritarianism and, too often, interfering with what Rock considered orderly progress. Rock clenched his teeth and prayed deliberately to heaven to deliver him from missionaries; after five days he sent Simpson back to Labrang.

The Tibetan escorts were nomads of the Sogwo Arig tribe, surly and disagreeable, but they brought Rock to Ragya without a serious accident. At elevations over 11,000 feet the countryside was bleak, still locked in winter, and botanically boring. The nomads along the way, though suspicious of the caravan, gave him no trouble. Curiosity got the better of their habitual aggressiveness. Rock paid the price in countless gulps of the inevitable rancid buttered tea. To amuse his motley hosts he took photographs and played opera records on his portable phonograph. The nomads called all foreigners Russians, or *Urussu*; the phonograph, thereby, became the "Russian magic box."

The expedition's real problems began at Ragya (*c.* lat. 35°, long. 101°), as Rock reported to Sargent on June 8. "After we got settled in our new quarters, we called on the *great* (?) Living Buddha of Ragya. He is a young man about 22 years and . . . claims to be the reincarnation of the *mother* of Tsongkapa. I delivered my letter to him from the god of Learning, the great Living Buddha of Labrang, and a few handsome presents and, after formal greetings, we began to talk business. That is, I asked his assistance in regard to our trip to the Amne Machin, seven or eight stages distant (by yak). His steward delivered a regular speech, for the Buddha seems to have little intelligence. He said that they are not on friendly terms with the Ngoloks as the latter claim that the Ragya lamas helped Ma Chi in fighting [them] in 1921. He suggested to make a quick dash to the Amne Machin ere the Ngoloks were aware of our presence, as they live some distance from the Amne Machin.

"To this I demurred as it was not simply my object to see the Amne Machin but to work there and I [would] rather secure the good will of the Ngoloks if that was at all possible. The nearest Ngolok tribe we had to reckon with was the Rimang tribe, and [the steward] said the chief of the Rimang is not at all a trustworthy fellow . . . he may promise to protect us and then arrange to have us robbed afterwards. I showed the Buddha the three letters I had addressed to the three tribes of the Ngoloks . . . given me last year by the Great Buddha of Labrang. We now discussed the possibility of getting these letters into the hands of these three chiefs. No one dared go as they were afraid of getting killed. After a few days searching for a possible messenger, one of the Living Buddhas (this monastery has fifteen), a Ngolok by birth, . . . was making ready for a visit to his tribe. The Ragya Buddha suggested to give the letters to him."[6]

Rock's wariness of the Ngoloks was historically justified. Other Western explorers in their vicinity, including W. W. Rockhill, Przewalski, and Abbé Huc, had depicted them as ruthless bandits, and none had traveled among them. One adventurer, Dutreuil de Rhins, though famous for his fearlessness, purposely chose a route to avoid the Ngoloks, only to die by other hands. Over the centuries, the Chinese had made occasional half-hearted thrusts into their territory, but always evacuated in haste.[7] Ma Chi-fu managed to collect taxes only by sending numerous armed troops who did their business and retreated quickly. Rock offered the tale of Chinese traders who had ventured into Ngolok country a week before his arrival in Ragya. They had killed a nomad thief and, in revenge, the tribe had descended upon their camp at night and made off with all their yaks. Instead of murdering the Chinese who, in their panic, had sent three horses, fifty taels of silver, and a rifle as peace offerings, the Ngoloks demanded life money of 4,000 taels, broadly suggesting that the traders had better pay. The latter bolted for Sining to beg General Ma's help. Prudence, Rock therefore observed, was of the utmost essence. While awaiting the replies of the Ngolok chiefs, he fled the Balkan-ish politics and wandered off to reconnoiter in the Yellow River gorges, *"absolutely terra incognita,"* he noted characteristically. It was, however, impossible to descend their sheer walls to the torrent below, so he peered down longingly and snapped pictures.

It was during one of these forays, on May 30, that he had his first clear view of the Amne Machin. Standing atop a high pass, he felt genuinely exhilarated for the first time in months. "I counted nine peaks, one a huge pyramid at least 28,000 ft. in height; it may prove higher than any Himalayan peak, including Everest. It is an enormous mountain mass towering above everything, and we were about 100 miles away as the crow flies." Nothing, now, could keep him from them.

But Rock's plans kept going awry. The letters from the Living Buddha of Labrang did not produce the desired results. Only one of the Ngolok chiefs even bothered to respond and he, presumably, was the least useful of them all. Instead of gaining safe passage, Rock had alerted the Ngoloks to his intentions. Meanwhile, one of the tribes had murdered a minor Living Buddha from Ragya near the Amne Machin, and the Ragya lamas organized a cursing expedition to the mountains to avenge his death. The steward of the lamasery suggested that Rock join the party, and he desperately agreed, provided he was furnished with an armed escort of thirty or forty Tibetans and some yaks. As neither man nor beast could be roused quickly, and the lamas were eager to get on with their cursing, Rock finally stayed behind. His efforts to arrange an alternative during the next week failed because no one in Ragya was willing to brave the Ngoloks with him.

Unwelcome as it was, the delay forced Rock to take a long look at Ragya and the crude existence of its denizens. "On the hillside back of the lamasery," he wrote later for the *National Geographic,* "are the huts of many hermit lamas, some even from the wild Ngoloks. Ceilings of these huts are so low that a man cannot stand erect in them; yet here these austere creatures spend their years. Others live in caves in the near-by cliffs. Prayer, meditation, and abstinence are their lot. They certainly *do* abstain! In winter they live on barley flour; in summer their chief diet seems to be stewed nettles."[8] "Appearances, however, were deceiving. The monkish inhabitants of Ragya did not restrict their ambitions to other-worldly concerns. A talkative old water carrier confirmed what Rock had always suspected: "the living Buddha business, he said, was a political or diplomatic system and always worked out for the good of the rich and influential. The local Buddha was very rich; so was his steward; and when "reincarnation" occurred, it seemed to him that this "miracle" always happened just as it might have been desired by the chief Buddha. For example, when the daughter of a powerful chief died, she was soon afterward incarnated in the person of a small boy, a nephew of the Budda's steward—a business and political arrangement agreeable to all concerned! When one of the minor Buddhas of [Ragya] died he, too, was happily, conveniently, and quickly reincarnated, this time in the person of the steward's brother!

"I smiled and asked the water-carrier how it happened that none of his children was the reincarnation of some departed Buddha. With a twinkle in his eye, he remarked that it was because the sum of all his worldly goods was two goats."[8]

Though such insights were diverting and provided future grist for his literary mill, they brought Rock no closer to his real objective. In desperation he finally resorted to blackmail. He insisted that a lama council be called. "I said unless they see me to the Amne Machin as they had first agreed, I would go to Sining and ask General Ma to give me a Moslem soldier escort and I would return to Ragya with them, and then the soldiers would make them get the number of Tibetans and animals, etc." The ruse worked, and two days later Rock was on his way. The horses swam across the Yellow River at Ragya while the men perched shakily on inflated goatskin rafts.

Not one single Ngolok showed his face between Ragya and the Amne Machin, thus leaving Rock in peace to do his work. Botanically speaking, the mountain range was a failure. Rock, always at altitudes above 15,000, collected seeds of a number of herbaceous alpine plants, but the slopes were bare of the trees and shrubs he sought.

It was also clear, upon closer inspection, that the principal peak fell substantially short of Mt. Everest. Viewing it from the valley floors, Rock could not make a precise reading and overestimated the height at around 28,000 feet. It is actually 23,490. Rock arrived at his figure with the aid of an aneroid barometer and inspiration—amateurish methods which would embarrass him later; but he was sufficiently confident of his readings to publish the altitude in the *National Geographic,* thereby unduly exciting the curiosity of a generation of mountaineers. The weather was good and the visibility excellent, so Rock made spectacular photographs; he also produced a detailed sketch map of the range.* Having accomplished as much as he could, there

---

* Albert H. Bumstead, the National Geographic Society cartographer, took a rather dim view of Rock's mapping techniques and encouraged him to improve them; but the late Erwin Raisz, also a professional cartographer, once remarked to me that he had seen Rock's maps and that they were of unusual quality for an "amateur."

seemed to be no reason to linger near the range and taunt the thus far invisible Ngoloks. He returned to Ragya in a week. No one had interfered with his party at all. His most alarming encounter, in fact, had been with two Tibetan women circumambulating the holy mountains by prostrating themselves inch-worm fashion—a process which they optimistically reckoned would take two months. Although he had not found a higher Everest, Rock was rather pleased with himself and the originality of his undertaking. He worried only that the bad botany would upset Sargent. The latter, however, grown philosophical in his old age, replied; "Do not forget that it is as important to discover that no plants grow in a country as it is to find what grows in it, and for this reason I consider your Tibet journey a success."

The return from Ragya to Labrang at the end of July turned out to be more hazardous than the trip through Ngolok territory because the caravan stepped into the middle of a fight between two nomad tribes. Rock, as usual, managed to parley and pay his way to safety, and made Labrang intact, in three months instead of the projected seven. By August 9 he was back in Choni. Prince Yang rode out to welcome him and deliver the latest bad news bulletins.

RAIDS AND TURMOIL IN KANSU

The political situation in the province had so deteriorated in Rock's absence that even little Choni, which he had presumed unassailable, mobilized for war. A serious crop failure in central Kansu had forced the prices of flour and barley sky high, and the Kuominchun continued to extort heavy taxes. The already restless population grew more excited. Three Chinese generals had combined to try to oust Feng's man, Liu Yu-fen, but the Kuominchun had beaten them back. The withdrawing soldiers vented their frustration upon the peasants in their path, burning, looting, and murdering as the spirit moved them. Liu declared martial law in Lanchow, and his army was entrenched around the capital. One of the defeated armies had retreated to a village near Choni, and it was feared that the Kuominchun would march upon them, hence Prince Yang's hasty preparations. There were rumblings among the Moslems, who observed the Chinese infighting with satisfaction; they had not, however, made their move yet.

Westerners were in danger from all sides. The people of Kansu, identifying Feng as the source of their misery, went on an anti-Christian rampage, attacking missionaries. Rock, as a white man, could easily be mistaken for one and, should the situation arise, would be hard pressed to describe his real business to an angry mob. He did not think this twisted situation was funny at all. He knew Feng was a communist and resented his claims to Christianity. Later, with Kuominchun encouragement, foreigners would be attacked as imperialists. A long delayed letter from the American Vice Consul in Kalgan, E. F. Stanton, stating that there had been considerable Soviet activity along the Sinkiang border and in Mongolia, fed Rock's fears that the "reds" were encircling the Chinese northwest and that Feng was party to a Soviet plot. Stanton requested Rock to write more letters describing events in Kansu as the United States had few reliable sources in that area.[9]

Turmoil in neighboring provinces spilled over Kansu's borders. Rock learned that Szechwan troops had crossed into Kansu at Pikou, the road he had taken, though no one seemed to know what they were up to. Guaita, the Postal Commissioner in Lanchow, wrote him that because of the fighting in Shensi between Feng's and Wu Pei-fu's forces, the road eastward was effectively closed. With the Szechwanese on the march, the route south to Chengtu via Pikou did not look promising either. The only avenue of evacuation seemed to be to the west, through Tebbu country, down to Sungpan in Szechwan and, eventually, out through Burma. But Rock could make no predictions because, to his absolute disgust, he still knew nothing about the money he had been demanding for months. "To talk about further plans *now* would be futile;" he wrote Sargent in August with heavy emphasis, "here, under the present circumstances, *one lives from day to day.*" Ammunition he had ordered from Shanghai had never arrived, and he was low on cartridges. He was also in danger of running out of provisions and grimly forecast that he might have to live like a native. "However, it might be worse, as the man said whom the devil carried off in a bag. The devil put him down in amazement and said 'what do you mean?' 'Well,' said the man, 'I might have had to carry you!'" So much for humor. "I can definitely say that if I ever get out alive," he told Sargent, "China will *see me no more*."

He grew morose and had nowhere to turn for moral support. Prince Yang, trembling in his furry boots, loaned him $800 and then buried all his cash; in his condition, the prince did not inspire confidence. Rock packed up some of the Nakhis and went off to collect in Tebbu country, establishing headquarters in a spooky old temple ("ugly masks with tongues two feet long . . . decorate the corridors" in a magnificent setting ("If the writer of Genesis had seen Tebbu country he would have made it the birthplace of Adam and Eve.") among wily nomads "Although the valley is inhabited by 'subjects' of the Choni Prince, they have declared their autonomy . . . never having heard of Wilson's 'self-determination' or of Wilson or even of America, for that matter. They form a sort of republic or a closed corporation, and woe unto the outsider who dares penetrate . . . their mountain fortress. The main object of their seclusiveness is to avoid paying taxes to an autocrat who gives nothing in return to his subjects save blows." The Tebbus of the Tara valley left Rock alone, the botanizing was heavenly, and he nearly relaxed for a few weeks. He even enjoyed a few laughs. Visiting familiar nomad encampments, he produced prints of photographs he had taken the year before and watched with amusement as individuals—many of whom had never seen so much as their reflections—recognized first each other and, eventually, themselves.

No sooner had he returned to Choni in late September than all vestiges of good humor disappeared. He was greeted by the news of the mixup over his money, and when he realized that Sargent had mailed his checks to the wrong bank, he exploded. He wrote a long, apoplectic, abusive letter, loaded with labyrinthian financial explications and threats that now, thanks to Sargent, he would surely never be able to get out of China. "While you are all enjoying good food," he noted pointedly, so as to fix the guilt in the proper quarters, "I cannot even get a cup of coffee or tea; native food is scarce owing to military commandeering of flour, etc. I am fed up on Chinese stuff such as this poor province furnishes. Rice is an unheard of article of food;

we have to grind our own flour and sift the chaff." Two days later he had cooled down enough to make back-handed amends for his rudeness, but he continued to be extremely agitated, very far from the nonchalant heroes packaged in Hollywood adventures.

Rock suffered indecision. Should he leave China or wait there? If leave, by what route? The Altai Mountains, Sinkiang, down through Tibet, and into India—a route once suggested by Sargent who was curious about the Altai flora—could be ruled out due to Soviet activities. Eastern and southern roads were hazardous. This left only an escape route through Tebbu country, and that filled with unknowns. He could, of course, wait, and hope that the chaos subsided even though, to date, waiting had never proved very worthwhile. But if wait, should he stay in Choni, now seemingly in danger, or should be look for another refuge among the nomads, where he would be absolutely cut off from information, where he would have to fend for himself and the Nakhis, perhaps through the winter? None of the options appealed to him, and he gave way to inertia, disregarding Sargent's entreaties to make for safety. He did not even go back to Tebbu country for further botanizing for fear of missing some important news. Instead, he sulked in Choni, listlessly packing crates of plants and birds and figuring schemes to get the packages back to Boston.

The longer Rock lingered, the worse things got. His enterprising old friend from Chengtu, General Yang Sen, had taken it upon himself to detain two British merchant vessels in Wanhsien on the Yangtze. A pair of British gunboats and an armed merchant ship, rushing to the scene, drew heavy Chinese fire resulting in the deaths of one commander, three officers, and about a dozen enlisted men, and a blot on the honor of His Majesty's Navy. Though the British failed to rescue their ships during the engagement, they fired into the town and set it ablaze. By the time Rock heard about the incident six weeks later the rumor had escalated: the British had killed 5,000 of Yang's troops and completely razed Wanhsien; all non-official foreigners had been ordered out of Szechwan; and Yang Sen had closed the Yangtze. Panic played havoc with reality. The destruction at Wanhsien had been exaggerated, foreigners in the province had neither been ordered nor advised to evacuate, and Yang had not closed the river. (U.S. authorities finally advised American nationals to leave Szechwan in January 1927.) But the episode had done nothing to endear Westerners to the Chinese. Meanwhile, Yang's provincial rivals had taken heart and attacked him, and trouble in Hankow had nearly brought Yangtze river traffic to a standstill.

Kuominchun troops poured into Kansu from Shensi every day. Having secured the Sian area at last, Feng decided to move more of his army and make Lanchow his headquarters. His forces fanned out into the countryside, collecting more taxes and, to use the current euphemism, "pacifying" the villagers and peasants. By the end of November Prince Yang had submitted to Feng's jurisdiction without even token resistance; what, after all, could a handful of soldiers and 3,000 armed peasants do against an army? Feng's American biographer, James Sheridan, cited the opinions of a writer in the *North China Herald* who described the extent of Kuominchun power in Kansu at the end of 1926: "First, . . . firm tactics have brought the entire province thoroughly under Kuominchun control; scarcely a dog dared bark against the Kuominchun. Second, Kuominchun rule brought heavy taxation, high living costs, and general dissatisfaction. Third, the bright expectations engendered by initial Kuominchun reforms remained largely unfulfilled."[10] Ma Chi and the Moslems simply lay low while unruly Tibetans, such as the Tebbus, remained law unto themselves. But Kuominchun rule was as close as Kansu had come to unified control since the Ming dynasty.

"Alas," Rock wrote Sargent, "I am sorry to report that even little Choni, situated on the edge of nowhere, has been ordered by Lanchow to hoist the Red flag. The Choni Prince had been ordered some time ago to prepare Red flags and now comes the order to hoist them, and so we are truly under a Red regime; but the Prince has as little love for the Reds as I have, yet he cannot refuse to obey orders or else he is merely driven out . . . China is really in a dreadful mess. The Reds (Canton) [Rock's term for the Nationalist regime in Canton—an uneasy coalition between Chiang Kai-shek forces and thoroughbred communists, such as Chou En-lai, which would rupture in 1927] have now appealed to the Lanchow governor to drive all Britishers out of this province, and I would not be at all surprised if he would consent to such a measure. . . . [November 14th] was Sun Yat-sen day in Lanchow, and students, girl students, Yamen employees, soldiers, and officials marched in parades and they yelled 'down with all foreigners, death to the British, hurrah for the Soviet-Kwangtung confederation.' A foreigner in Lanchow wrote me that it resembled the revolutionary days in Moscow of 1918." Within less than a week of this letter to Sargent, communist propagandists had moved into Choni, shouting anti-foreign slogans in the street. People who had been friendly the day before stopped speaking to Rock. The price of hay for his horses doubled. The Lanchow authorities spread word that they would not abide by the laws of extrality.

Sargent pleaded with Rock to quit China, but the explorer always countered with some excuse. By December he had managed to straighten out his financial problems more or less satisfactorily. He was, therefore, left with political disturbances and, as winter closed in, the weather, as obstacles to his exit; he was also fond of citing his obligations to the Nakhis who, one suspects, felt even more responsible for him and, had they been left to their own devices and without the promise of a bonus, would have hurried back to their families in Likiang, whom they had not seen for two years. Rock's diaries and letters written during the winter of 1926–27 gave the impression that he was trapped in Choni, yet his mail was getting through, and he received letters from Boston in about two months. All the boxes of birds, seeds, and plants specimens that he mailed from Choni to Shanghai for shipment to the United States got through; he knew this because the Shanghai postoffice sent him receipts. Due to his meticulous attention, not one shipment of specimens went astray, and all of his letters reached Boston, however delayed. Rock was, in fact, as much a captive of his own fascination and fear as he was of China's disorders.

He watched the installation of a new living Buddha in Choni, "a dirty, unwashed kid," aged 5.* Mostly to test himself,

---

* This particular living Buddha, one Zemelin, came to a sorry end. In 1948 he was defrocked near Lhasa because he kept a woman; soon after he died mysteriously from poisoning. Rock learned about Zemelin's fate when he was in Darjeeling in the early 1950s and thoughtfully added the note to his diaries.

he revisited the demonic dances at Hochiassu, succeeded this time with his photography, and suffered only a queasy stomach. He observed another butter festival and noted, as signs of decline in Choni, it was far inferior to that of the previous year; one sorry looking butter god looked as though he had been held over from last year and dragged out for a second performance. Rock arranged the purchase for the Library of Congress of a set of wooden printing blocks for the Tibetan holy books, the Tandjur (commentary translations) and Kandjur (canon translations). But there were few novelties for him in Choni during this second winter.

Time passed slowly and disagreeably. Simpson rode down from Labrang occasionally to feed his anxieties with gossip about the Moslems. The mission in Choni had passed into the hands of two pleasant maiden ladies, Misses Hansen and Alquist, and Rock stopped for tea now and then, but missionaries, in principle, made him nervous. The citizens of Choni became increasingly bold, taunting him with their freshly learned anti-foreign slogans. Prince Yang, though always friendly, was too preoccupied with his own problems to be good company. The Nakhis were restless and uncharacteristically quarrelsome. By the end of January Rock realized he would go mad in Choni and suddenly started preparations for leaving, as though ice, snow, bandits, and soldiers had magically vanished.

The usual delays ensued; mules, men, and provisions being more difficult than ever to obtain. Rock did not press too hard, so it was March 10 before he actually made his formal departure. Yang sold him extra rifles. Rock made a small "loan" to the missionary ladies so they would not be stranded. On the morning of the 10th, the priests of the lamasery flocked to his temple quarters in the early morning, pretending to say goodby, to scavenge the debris he had left behind: they stuffed obsolete issues of the *Atlantic Monthly* and the *North China Herald*, bottles, pins, screws, handles, used razor blades, and scraps of paper under their robes and retreated ecstatically. Prince Yang waited for Rock on horseback in a grove of poplars beyond the city gates and bid a quick, manly farewell, doffing his hat as he rode away. The gesture touched Rock deeply, and he swallowed hard. The march out of China was on at last but, sadly, the person who had worried most about Rock, Sargent, died before receiving the news.

Rock took the route through Tebbu country to Sungpan in Szechwan, and the trip started badly. Six inches of fresh snow fell during the first night, and it kept snowing through the next twenty-four hours. The temperature dropped to 18°F., and the wind blew five foot drifts. Rock and a few Nakhis plunged ahead, but the caravan stuck in the snow. The Tebbus refused Rock shelter and food, so he had to camp outdoors.

On the fifth day, "we saw my cook galloping alone and yelling, so we turned to meet him. In a most exhausted condition he told us that our caravan was being attacked by 30 Tebbus . . . We turned immediately and galloped as fast as our horses could carry us. The Tebbus had ambushed the caravan. They knew me, of course, but they did not think that the caravan belonged to me but to . . . Moslem traders as I had never come to the Tebbu country with such a large caravan. Fortunately they had no rifles but attacked my men and muletiers with huge sticks and rocks. My cook suffered most. They grabbed his horse and one man hurled a large rock into his chest. While another was in the act of laying low with a huge stick, my hunter saved the cook's life by emptying a no. 6 shell from the Marlin shotgun at 2 ft. distance into the would-be assassin's legs; the force tore his garment to pieces and also his thighs. The escort fired, killing two Tebbus at once. One was shot in the head, and another in the chest. Another Tebbu had his jaw knocked out by a husky muletier. We lost nothing, they fled up the hillside, but yelled that they would fetch their rifles. . . . They are a villainous bunch. I now wonder how we escaped for the last two years the clutches of these wild wretches." Afraid that the Tebbus would attack again to avenge their dead, Rock ordered a furious pace and spent a few sleepless nights.

The snow lay so deep that the caravan had to dig its way through the high pass at Yangpu shan on the Kansu-Szechwan border. They surprised another gang of bandits—the Tebbus having abandoned the chase—and managed to disarm them efficiently. After eighteen exhausting days on the trail, Rock marched apprehensively into Sungpan.

Much to his surprise the officials were ready for him, courteous and obliging. Rock had written Yang Sen in Chengtu, and the latter had dispatched orders that Rock was to be protected. Because of the Wanhsien incident, anti-British sentiment ran high, but the citizens of Sungpan were still friendly toward Americans. Rock was told that all but one Englishman had left the province and that other Westerners, fearing that the Chinese would become less discriminating in their hatred, were rapidly following suit. But Yang's cooperation encouraged Rock and, instead of taking the route through the mountains into Burma, he decided to travel to Chengtu and thence to Yunnanfu.

Once outside Sungpan, however, Rock was harassed by jeering mobs in nearly every town. He behaved stoically, passing through them without even so much as a twitch of a facial nerve while inside he seethed with rage and fear. The concentration required for composure enervated him more than watching for bandits, and he frankly wished he had taken the mountain road. To escape insults, he now often traveled in his sedan chair with the curtain drawn. On April 14 he arrived in Chengtu, emotionally spent; only ten days earlier the *Washington Post* had written: "Harvard Explorer Is Missing in Tibet." A few Americans connected with the University and some missionaries had elected to stay behind in Szechwan. Students had tried, with partial success, to starve the foreign faculty out of Chengtu, but a few tenacious souls carried on and welcomed Rock in their midst. The Chinese military authorities were particularly helpful—Yang's influence at work—and suggested that, since the Japanese consulate was folding its tents, Rock might travel to Shanghai via the river with their party. The Japanese agreed to the arrangement, and Rock told ten of his Nakhis to return to Yunnanfu alone, his presence among them having become dangerous to them. The other, his cook, he kept for his journey to the coast. The twelfth had been sent back to Likiang after the opium incident. Technically the expedition had ended. Rock had only to leave China now.

The strain of the last two years finally exacted its toll. Rock, having surrendered himself to the charge of others, collapsed physically and mentally. On the way through the Red River basin in Chungking he fell ill. "I am so disgusted with this country and the whole journey," he wrote, "that I do not care at all about writing down the names of places . . ." He gave up diaries and letters and, between Chungking and the coast, spent most of his time on a stretcher, thus becoming notorious as the man who "had to be carried out of China in 1927." Though the river journey was not without its vicissitudes, the party arrived safely in Shanghai.

There Rock caused his protectors and the American authorities no small amount of consternation by recovering his strength and insisting upon his independence, wandering about on his own outside the International Settlement. He immediately got fed up with the "so-called civilized life," went to Hong Kong for a week and then headed back to Yunnanfu to reestablish himself and make certain the Nakhis were all safe. His eccentric perambulations further annoyed U. S. officials, who were making every conceivable effort to concentrate American nationals in port cities where they could be guarded. As it happened, four of the Nakhis had been sick on the way to Yunnanfu, and Rock found them recovering in the mission hospital. He stayed with them until he felt sure they could travel back safely to the homes from which he had summoned them so long ago. Necessary or not, his generosity touched the men and goes a long way toward explaining their loyalty to him in later years. Having satisfied his conscience and gathered all his belongings, Rock left China only out of necessity. He headed for the United States in mid-summer of 1927. He had two Nakhi men with him and he planned to return.

Behind him lay an embattled nation. Feng Yu-hsiang had quit the northwest during the spring, first cooperating with communist forces masterminded by Borodin and, later, after Borodin had gone back to Russia, tail between legs, and the Chinese communists had been expelled from the Kuomintang, committing his forces to Chiang Kai-shek's Northern Expedition and indicating that he favored opportunity over ideology. Feng's withdrawal from Lanchow created a vacuum that made the Moslems bold again. A Moslem rebellion broke out in 1928, true to Rock's predictions. Though its details were obscured by more vigorous and, to the West, more pertinent, fighting in the east and south of China, the death rate in Kansu was in the hundreds of thousands. Yang Sen of Szechwan threw in his lot with Chiang

Kai-shek. Lung Yun, newly arrived governor of Yunnan, fought a few battles against anti-Kuomingtang warlords in the southern provinces on Chiang's behalf, strictly, however, for the sake of the spoils, and concentrated his talents on private corruption. Chiang launched his unification campaigns in the nouth; he was China's rising star. But, while he subdued the northern warlords, he made a formidable new enemy out of the communists, and they would haunt him for the rest of his life.

VII

CRUELTY AND INEFFICIENCY OF THE RULERS OF CHONI, MULI, AND YUNGNING

Before 1927 ended, Rock was back in China as an explorer for the National Geographic Society and, up until 1949, neither civil war nor Japanese invasion could keep him out; only the present communist government succeeded in that. He never returned to Kansu but spent most of the time exploring in southwest Szechwan or northwest Yunnan and, increasingly, turned his attentions to the language and history of the Nakhi tribe. The last time he saw the prince of Choni, therefore, was March 10, 1927, when they parted beneath the poplars outside the walled lamasery. However, on his next expedition, Rock renewed his acquaintances with the king of Muli and the Tsung-kuan of Yungning, thus keeping intact his ties with royalty—or what passed for royalty along the Tibetan borderland. The uniqueness of each principality attracted him, but more compelling to someone as interested in power as Rock was the proximity to power: its exercise, its compromise, and, as it happened, its decline.

Rock always referred to Choni, Muli, and Yungning as semi-independent kingdoms, adopting kingdoms as an approximate analogy. It was a peculiarity of Chinese bureaucracy that, while all three potentates held the power of life and death over the peoples in their respective territories, not one of them could claim legal sovereignty. Each one was beholden, to a greater or lesser extent, to a district magistrate who, in turn, paid allegiance to a provincial governor, and so on up. Under the Manchus the emperor had been the ultimate authority; during the free-for-all of the Republic, authority depended on force.[1] In principle, Chinese administration was uniform throughout the land, but in marginal regions inhabited by "barbarians," the dynasties, much less the Republic, never had secured control. It was considered inexpedient to dispatch the big armies and occupying forces needed to bring the barbarians permanently to heel. The dynastic solution had been to push them off the arable land, govern those who, like the Nakhi, permitted themselves to be governed, and let the rest govern themselves upon payment of tribute. The arrangement suited the Chinese who, interested in taxes and trade routes, feared and despised the tribespeople. It was thus that such enclaves as Choni, Muli, and Yungning, geographically remote, maintained their semi-independent status into the twentieth century, and their chiefs, called *t'ussus*, played whimsically with power for the edification of a Western explorer.

CHONI OF THE TEBBU

Rock was not the first Westerner to visit Choni any more than he was the first in either Muli or Yungning, but he was the first non-missionary to spend any appreciable time in any one of the three. In fact, other botanists had passed through Choni. Grigori Nicolaevich Potanin, an enterprising Russian, stopped there during his third expedition in 1884 according to Emil Bretschneider's *European Botanical Discoveries in China*. His wife, who used to help him collect and prepare botanical specimens, was in the party, an arrangement which made Rock shudder every time he thought about it. Reginald Farrar visited in 1914. It was General Pereira however, who originally brought Choni to Rock's attention as an ideal launching spot for an expedition to the Amne Machin.

Choni, a district of some 64,000 inhabitants, nestled in the extreme southwest of Kansu, bounded on the north by the *hsien* cities of old and new Taochow, on the west by Tibetan grass-lands, and on the south and east by unadministered parts of Szechwan. The principality originally had been established during the glory of the Ming dynasty at the beginning of the 15th century. Five centuries later the boundaries had become fluid, subject to what the prince was not prevented by rivals from controling at a given moment. He ruled tribes and families rather than a defined land area. Ironically, while most of Kansu was flat and arid and, therefore, unfertile and boring, little Choni, with its limestone mountains, alpine meadows, forested valleys, and two rivers, the Tao and the Pailung Chiang, was a scenic marvel and—had anyone thought to take advantage of them—rich in forest products. Potential or no potential, it was the craggy kind of landscape which did not attract the agriculturally-minded Chinese. The 10,000 Tebbu tribesmen who occupied the western and largest area of the principality, used the forest destructively and stupidly.

Rock encountered his first Tebbus in the Yamen at Choni. In describing "The Land of the Tebbus" in the *Geographical Journal* he wrote: "There were six of them all chained together by their necks. They were from lower [*i.e.* southern] Tebbu Land, tall gaunt-looking dark men like Hungarian gypsies. When I inquired if they were robbers, their guards said 'Oh no.' On further inquiry I learned that I was indirectly responsible for their plight. Some months previously I had given to the Choni prince seeds of vegetables and melons which the Department of Agriculture of Washington had sent me. The melon seeds he sent to lower Tebbu country to be planted, as the climate is much milder, and these poor wretches were to take care of them, and when ripe to bring them to Choni. Having never seen a melon they allowed them almost to rot on the vine ere they sent them to the Choni prince. This enraged the feudal lord, and forthwith he had them brought to Choni in chains to have them punished for their seeming negligence."

According to Rock, who spent many weeks collecting in their territory while trying to make up his mind to go somewhere else, the Tebbus were without competition as the stupidest and wildest of the Choni prince's charges. "Every village is at feud with the other, and at every approach there is a wall with loopholes . . . he that ventures into Tebbu land, let him beware; it is a robber's den inhabited by cruel, revengeful, suspicious people, who plow with loaded rifles and carry long swords with hand always on the hilt when talking with a stranger or even a neighbor, for none is certain what [course] the conversation might take." In contrast with lower Tebbus and most Tibetans who were around six feet tall, Rock remarked that the upper Tebbus were undersized; he also noted a high frequency of cretinism. Rock believed the Tebbu tribe to be a branch of the ancient Chiang tribe from which the Nakhi and Moso had also descended. Owing to their geographic isolation, the Tebbus had remained ethnically quite pure. Whatever they had been in the past, relations between the Tebbus and the Choni princes were not at all amicable during Rock's stay, as the melon affair indicated. Since Yang did absolutely nothing for them, the Tebbus hated him. He could not call them to arms when he needed soldiers and he could not pass safely through their territory without a huge armed escort. Rock considered it only a matter of time before the Tebbus, the largest single tribe among the prince's subjects, organized some kind of rebellion and told him to mind his own melons. However, internal squabbling weakened their resistance to his predatory policies. The prince, meanwhile, drew his militia from a more civilized tribe of Tibetan stock, the Choni-tzu, who lived in the Tao valley.

The seat of power was the village of Choni itself. Pereira described it as a town of around 320 families, of whom about half were Chinese and the other half of Tibetan stock, though few pure Tibetans, while Rock claimed it had 2,000 inhabitants of whom there were very few Chinese. Regarding the ethnic complexion, one must follow Rock, who spent two winters there, rather than the General, who compiled his observations in a few days in March, 1922. A lamasery of the reformed Yellow Sect, called Chantingssu or the Monastery of Abstract Contemplation, stood 500 feet above the village proper and was composed of 172 buildings and ten chanting halls. During the Yung-lo period of the Ming dynasty, some 3,800 lamas had occupied the lamasery but—signs of spiritual decay—Rock counted a mere 700. The American-based Christian and Missionary Alliance maintained a mission in Choni, with little effect. The Tibetans of southwest Kansu were no more susceptible to Christianity than the Nakhi of Likiang. For two winters Rock camped in the carved, painted, and lacquered home of an incarnation who had come from Lhasa to Choni and remained there to die. The lamasery, ancient and in remarkable condition considering the earthquakes which shook the mountains, had beauty and dignity. The village below was squalid.

Rock, finnicky and disposed to indulging in gory details, observed with horror that the people only cleaned their homes once a year, shortly before the Chinese New Year. "The streets were littered with filth which had been swept out of the houses; it remains on the street to show that the houses had been swept. The dirt in the main street is disgusting, to say the least. The pigs are having a grand time in the mess . . . " The annual cleaning was a common Chinese ritual made unsanitary in Choni by the absence of a direct water source with which to wash the streets. Water was supplied by the Tao River, which flowed inconveniently a third of a mile below the village. Women carried the water up to the village in wooden buckets; the poorer lamas transported it the additional 500 feet up to the lamasery. The water went for cooking and drinking, not, Rock noted caustically, for bathing. Both residents and lamas reeked of rancid yak butter, and their skins had become black with an accumulation of grease and dirt. Only the highest lama officials cleaned their faces, which produced lighter features set in the darkness of their unwashed bodies and made them look peculiar to a Westerner. Rock refused to be impressed by the argument that the butter and dirt layer protected the poorly dressed people against the winter.

The mountain Tibetans were by tradition nomads, not farmers, and even permanent settlements like Choni were agriculturally dependent. The village and river valley people raised livestock and vegetables but bartered for grains grown on the Kansu plains; the nomadic Tebbus raised sheep. When the price of horsebeans and barley skyrocketed, as it did during the winter of 1926-27, Choni found itself on the threshold of starvation. In the tiny town, the 2,000 inhabitants managed little more than subsistence living. Hundreds of people apparently did even less, and the streets were filled with urchins and destitutes begging coins for food, fuel, or opium. And thus life had gone on for centuries, an essentially peaceful but mean existence interrupted only by the excitement of religious festivals, a life in which the hardships of securing necessities generated a pettiness of spirit far from the idyllic fantasy of Westerners' imaginations. Choni was no Shangi-la. Only the natural setting lived up to dreams, but it was hard to admire the scenery on an empty stomach.

PRINCE YANG OF CHONI

This collection of monks, villagers, and surly tribesmen was ruled by the Prince of Choni who was, of course, only a prince in translation; in fact his title was *t'ussu*—headman or ruler. Under the rules of succession, if a prince had two sons the elder succeeded him while the younger became the grand lama; Prince Yang Chi-ching, having been an only son, held both offices. In the early days of the *t'ussu*-ship the ruler had used his Tibetan name but later was asked by Peking if he would like a Chinese name. The conferring of Chinese names and honorific titles upon "barbarian" potentates was a stroke of Chinese diplomatic genuis, obligating the recipient toward his benefactors and, under the guise of flowery words, weakening his position. The Chinese played successfully upon human vanity. This system continued into the Republic, perpetuated more by the greed of provincial authorities than by any directives from the embattled governments in the east which did not have time to formulate any "barbarian policy." While the relationship between the provincial governor in Lanchow and the central government was tenous at best, that between the governor and Prince Yang of Choini operated as it had in Ming and Manchu times.

Pereira had thought Yang a nice fellow, friendly to foreigners; Rock, while finding him hospitable, saw too many samples of his cruelty to forgive him. Yang, part Tibetan, part Chinese, was in his mid-thirties when Rock met him. He was the only well-dressed person in Choni. He wore Chinese silks and satins on ordinary days and, for festivities, donned a Tibetan costume with fox hat and stylish fur boots with upturned toes and multicolored leather stripes.* Middle-sized, slender, and clever, he had none of the Muli king's childish naivete. In many respects, Yang was quite a modern man with a surprisingly good understanding of the outside world. He did not pose foolish questions and he listened intelligently. He did not make his subjects grovel before him but, since he was known to chop off an ear for so slight a misdemeanor as not responding instantly to his call, they grovelled voluntarily.

Rock diverted Yang from his humdrum power problems. The prince did everything to make the foreigner comfortable and treated him with a deference he had never bothered to show the missionaries. He helped Rock fetch men and mules for expeditions, wrote letters of introduction for him, sold him guns, loaned him money, and took, or at least pretended to take, more than a passing interest in his botanical collections.

A holiday banquet in Rock's honor was a touching, if unappetizing, example of his good intentions. "He tried to make it half foreign, half Chinese," Rock wrote afterwards. 'The result was an utter failure. In the middle of the meal he served watery ice-cream with a flavor reminding of cheap toothpaste. The ice-cream was served with the gelatinous, slippery tentacles of squid . . . out of a tin can. Then melon jelly with seaweed and carrion pork. A big bottle fly was buzzing about, and a Chinaman present caught it with his chopsticks and then tried to help me with the same chopsticks to doubtful morsels out of an unclean bowl. Beer was served—Japanese lager beer—in tiny silver cups such as are used for liquors. And this with ice-cream and devil-fish. I was glad when the whole thing was over and wished I could have been spared the little I was forced to eat in order to be polite."

Rock feasted upon the flattery, if not the food, of the Prince's attentions, but grew uneasy and irritable as he collected evidence of the man's meanness. Since Yang was only the grand lama by heredity and had never taken religious vows, he was not bound by rules of celibacy like ordinary lamas. Accordingly, he led a lusty life with four wives and an assortment of slave girls who doubled as concubines. He treated his women very badly.

---

* Rock posed in a similar costume in a photograph of which he made many prints; almost everyone who knew him well had a copy.

"His wives, slaves, girls, are often beaten or tortured. Some have disappeared, having been killed in the Yamen, and news is given out that such-and-such a slave ran away." Rock knew what he was talking about because he had nearly become involved in one incident. "One morning a lama brought a middle-aged woman to my quarters in the lamasery. He handed me a present of cakes which had been brought by the lady who went on her knees and kowtowed before me. I urged her to rise and inquired of the lama who she was and what she wanted. The lama told me that she was the wife or *Tai-tai* of the Prince and wished me to intercede with her husband for her. She had an aged mother who lived in Hochou and whom she wanted to visit, but the Prince refused to let her go. She was certain, she said, that if I asked him to let her go . . . he would do so. I gave her her box of cakes, but as sorry as I was for her I told her that I could not interfere in the family affairs of the Prince, and she left with tears running down her cheeks."

The woman, however, did not give up and disappeared a few days later. Because her feet were bound and she was unable to walk—the strategic value of foot-binding to the possessive husband being amply demonstrated in this episode—she had to hire a muletier to carry her on his back to Taochow, where she wished to hire a chair. As soon as the prince realized she was missing, he dispatched runners to all exit routes, and they found her on the muletier's back. Following the prince's orders, her captors tied her hands and feet to a long pole and carried her back to Choni swinging like a pig. Rock heard that she was beaten savagely. The next day she hanged herself from a rafter in her room.

Yang's domestic difficulties did not end with his women. His only son and heir smoked opium. Yang himself was a smoker, but deliberately disciplined his consumption. The son, however, was heavily addicted in his teens and could not hold a teacup steady. The prince, alarmed by the boy's condition, posted a proclamation that anyone caught selling opium to his son would be flogged. In desperation the boy found a six-year-old child, gave him money, and threatened him with a beating to secure the drug for him. Yang finally forced his son into revealing his supply and sent men out to punish the child—the idea of punishing his son seemingly eluding him. The child was stripped naked and tied, Christ-like, to a cross in a field. There he lay in the middle of winter where he would have frozen to death had not Rock interceded on his behalf.

Yang's principal administrative organ was a corps of personal servants and runners who spied on everybody and, no doubt, on each other. An opportunistic lot, they grew rich on threats and bribes but flaunted their loyalty to the prince by adopting his name; one of them was called, appropriately, Yang T'ou-mu, or Yang-Eye-in-the-Head. The people despised the Tibetan *Ton-ton macoutes* to whom justice was an unknown concept. They were the culprits who reported to Yang when his melons and squashes were neglected, caused the Tebbus to be chained and tortured, and tracked down the wayward wife. It was, then, not without a degree of satisfaction, that Rock noted many missing ears among them.

From the subjects' point of view, the prince lacked lovable qualities. He extorted fines from them to fill his coffers and pay off his Chinese oppressors in Taochow and Lanchow; he sent his spies to watch them and his soldiers to beat them; he offered nothing in return. He was, accordingly, most respectfully unloved. Yet, in the eyes of his people, Yang represented a buffer between them and the even more hated Chinese. They did not comprehend that the Chinese had, over the centuries, made Yang the instrument of their collection. Absorbed in his vanity and power, not even Yang could appreciate this shabby truth. When he paid squeeze money to some provincial official, he imagined he enhanced his power while, in fact, he only proved his efficiency as a factotum. As a result, the Chinese used any flimsy pretext to demand more money, and Yang was harrassed on all sides. The more the Chinese backed him into a corner, the more pathetically he flailed.

At least one faction among the presumably dim-witted Tebbus was clever enough to realize that, though lacking in power themselves, they could make Yang squirm by appealing to his Chinese superiors. In his diary Rock made note of an illuminating story: "The Tebbus of Kadjakin [valley] lodged a complaint against [Yang] in Lanchow with the Governor. The latter sent a deputy to investigate the affair. The prince naturally greased the deputy's palm—and handsomely—in order that he give a good report about him. The [Taochow] magistrate, under whom the prince rules or to whom he is subordinate, had to have also a handsome squeeze, so the prince gave him $3,000, and between the Lanchow deputy and the . . . magistrate, the affair was settled. It happened, however, that the city magistrate, just a few days after he had received the $3,000, lost his position and was replaced by another magistrate. The prince called for volunteers among his underlings to go to [Taochow] with Choni soldiers and prevent the old magistrate—who had lost his job and consequently his authority—from leaving . . . until he had returned the $3,000. One fellow volunteered and, with a few soldiers, went . . . and camped in front of the magistrate's place, which, in fact, belonged to the prince, the latter allowing the magistrate to live there after he had lost his position. One day the old magistrate said 'we are going,' and opened the door; the delegate said 'oh no you are not until you pay back the squeeze money.' His, the magistrate's, escorts drew their revolvers, . . . the Choni men were afraid, and the old magistrate decamped with the $3,000."

When they were not after Yang for money, the Chinese demanded soldiers. Yang could muster forces of around 2,000 men whom he dispatched at the pleasure of the governor in Lanchow or the Taochow magistrate to fight Szechwanese bandits, marauding Moslems, or competing Chinese. As provincial politics became more confused, the Chinese increased their demands, and Yang grew nervous. The Feng Yu-hsiang regime made him especially anxious, upsetting the familiar order and encroaching upon his territory with their banners and slogans. The hoisting of the Red flag in Choni, an act that Yang resisted as long as he thought he could get away with it, humiliated the prince far more than the loss of the $3,000 to the old magistrate. He interpreted it as a bad omen, hurriedly stashed his gold, and—for what eventually were revealed as self-interested motives—proclaimed eternal friendship for Rock. Yang's pessimism proved realistic. Some months after Rock's departure he learned from a missionary in Kansu that Feng's forces had stripped the prince of his title, confiscated his domain, and retained him to govern it as "commissioner to the barbarians," entirely at their mercy.[2] Clearly,

Feng was no more prepared than other Chinese to take on the Tebbus, and Yang's demotion made very little practical difference. The change of status was, however, a dreadful loss of face, and the prince was frantic because there was no authority to whom he could appeal his case. In comic desperation he wrote his friend Rock, begging for help. The latter did not respond. "I had feared such a fate would befall the poor man sooner or later," Rock remarked, with momentary sympathy for the man he knew was vicious.

Yang, however, survived the damage to his dignity and advanced to other troubles. When Feng's army evacuated Kansu in mid-1927 to join Chiang Kai-shek on his 2nd Northern Expedition, it left a vacuum. The old Moslem-Tibetan feud, quiescent during Feng's tenure, revived in earnest and, during the spring of 1928, reached Choni. According to the version that reached Rock, Yang's soldiers captured four wives and a treasury belonging to one of General Ma's sons. The Choni soldiers delivered their prizes to Yang who absconded with the women into Tebbu country, where he raped and murdered them. In revenge, the Moslems sacked Choni, burning the lamasery, including Rock's old domicile. The Tibetan population was slaughtered and the missionary who described the scene to Rock said that corpses littered the streets of Choni and cholera broke out. The Moslems, however, would not pick a fight with the Tebbus, so Yang and his family slipped away with bodyguards and hid among the tribesmen. After the Moslems left, he returned to Choni and rebuilt the Yamen and his summer residence.

Ultimately, with poetic justice, it was neither the Chinese nor the Moslems who brought about the Choni prince's downfall, but his own subjects. One night he heard machine gun fire outside and, thinking the Moslems had returned, escaped through a small door in the rear of his Yamen and hid along a stream bank. When he recognized the voices of his servants, he revealed himself, believing he was rescued. It was a natural, but fatal, miscalculation. "They bound his hands behind his back, bound his feet, and threw him into the rocky, shallow streambed. . . . Then they dragged him up and down over the rocks until he was dead. Thus came to an end a cruel autocrat."

The coup had been engineered by an officer. All of Yang's family except his son and one wife died before machine guns.

The Chinese, to whom these matters were reported, wanted to put a magistrate in Yang's place but, ironically, the Tebbus, who had always despised him, went in mass to Choni and protested. Yang might have been bad, but any Chinese would be worse! The Chinese, realizing the hopelessness of trying to cope with the Tebbus, relented and appointed Yang's son, the opium addict, "Chief" of Choni. In his condition, he undoubtedly made an even better lackey than his father had been.

THE RULER OF MULI

When Rock was plotting his course out of China in the spring of 1927, he had planned to travel south through the mountains of western Szechwan, down to Muli, and thence to Yungning and Likiang. But, owing to the warmth of his reception in Sungpan and the promise of safe conduct, he altered his itinerary and went to Chengtu. Back in the United States he persuaded the National Geographic Society to sponsor his exploration of unmapped mountains in western Szechwan. Muli was selected as a convenient base for this expedition as Choni had been for the Amne Machin party, and Rock arrived there on May 26, 1928, after an absence of more than four years. Chote Chaba, the Muli *t'ussu* whom Rock called 'king', was positively delighted to see him again and welcomed him with open arms to his domain.

Muli was even more impregnable than Choni; geographically, the *t'ussu* had the upper hand over the Chinese. Roughly 9,000 square miles in area, slightly larger than the state of Massachusetts, it was jammed with mountains and rivers and a few stretches of level land. The valley floors started at 7,000 feet, and Rock guessed that the average mountain altitude was around 14,000. A territory both difficult to enter and to traverse, as well as surrounded by bandits and Chinese-hating tribes, it afforded its ruler good protection from the advances of the Chinese and almost complete practical sovereignty. "Nowadays," Frank Kingdon Ward announced after his visit there in 1922, "Muli pays tribute to no one."[3] Muli even appears to have been impervious to Christian missionaries, for Rock never encountered any there. The principality supported a lama population of 12,000 and tribesmen no one had bothered to count. Ethnically, Muli was a hodge-podge of "barbarians." There were Hsifan, some Nakhis, Moso—closely related to the Nakhis of Likiang—Lolos, and several other peoples, speaking different languages and dialects and all more or less mistrusting each other.

Muli had been under the jurisdiction of Yungning until the 17th century. At that time an enterprising lama arrived from Tibet and converted the tribes from Black Shamanism—the Bon, or pre-Buddhistic religion of Tibet—to the Reformed Yellow Sect of Buddhism.* As a sign of his gratitude and an inducement to remain in Szechwan, the t'ussu of Yungning awarded the lama all the territory to the north of his district. This lama became incarnate in a Tibetan after his death; in 1724 the incarnation journeyed to Peking to pay tribute to the emperor Yung-cheng. The emperor wished to appoint him t'ussu of Muli, but the lama begged off, claiming that as an incarnation he was not of this world. He suggested that the emperor name instead his steward, the descendent of a Manchu who traveled with Kublai Khan, and Yung-cheng obliged him. The precedent for a theocracy was established and, from that time, the t'ussu was always a lama and the second son in the family; lamas being theoretically celibate, the oldest son carried on the family. The Manchu name long had been discarded in favor of a Tibetan name; the Chinese further endowed the family with the name of Hsiang.

Kingdon Ward had known Chote Chaba's predecessor, the Muli king who had once warned Rock off a journey, quite well, as a "tall, corpulent man of about sixty," who was "not insensible to the advantages of progress." His Tibetan subjects being poor and unskilled, capable only of fashioning tsamba bowls, he imported a number of Chinese into the territory to pursue more exalted crafts such as carpentry, cobbelry, etc. He gave the Chinese free lodging in the monasteries and paid them by the job. In this respect, he appears to have been an enlightened rul-

---

* The lama's efforts were not entirely successful. Rock reported a pocket of Shamanism in the Tsoso district.

er. He was, however, extremely jealous of his power and carefully selected as his "prime minister" a cheerful young man who was totally illiterate. When the king died in 1923, his nephew and heir was still a minor, so the lamas of Muli decided to call on one of the king's younger brothers, also a lama, to rule until the nephew became of age. The Chinese, one notes, were not at all involved in the selection.

Perhaps corpulency was a genetic trait, but Rock thought it was an occupational hazard. "All Muli kings were excessively fat from lack of exercise, sitting in meditation, and eating well, especially butter and much sour cream with plenty of tsamba . . . and rich Chinese food, but mainly from drinking 20–40 cups of buttered tea between meals." Chote Chaba, a mountainous man, did honor to the royal tradition. His "only exercise is clapping his hands when he wants one of his slaves to approach him." On second encounter, Rock found him fully as genial and more rotund than he had been on the first. Seldom having set foot outside his principality, he was as gluttonous for information as for buttered tea. Time had not improved the quality of his questions; he was still grotesquely innocent. "News had traveled fast," Rock remembered, "and the king had heard that I took two men with me to America, also that they had flown over Washington in a sight-seeing plane. He called these two men into his presence, and they related to him the marvels they had seen. He asked them how long it took for a boat to cross from Shanghai to Chin shan (gold mountain, as California is known in China), and when they replied twenty days, he raised his eyebrows and exclaimed 'Aleh! What a long rope they must have to pull the boat across.' [This by analogy to the river ferries in Muli, which were simple dugouts fastened by a ring to a rope and towed across the river by pulleys.] He had also heard that foreign countries kept wild animals in enclosures like those of the Dalai Lama in Lhasa where, on his one and only visit, he had seen a tiger and an elephant; he inquired if they had any dragons in the Washington Zoo. He wanted to be told about the sensation of flying and if the wings moved like that of a bird and, as a climax, he asked 'When you were high over Washington, *could you see China?*'" Rock fielded such queries with dexterity. He and Chote Chaba were a curious sight together. The lamas, even the higher incarnations, were forbidden to sit in the king's presence; the peasants were not allowed to gaze upon his person; but the white explorer sat with him for hours, swapping information. Rock learned of strange customs.

"I was invited to have lunch with the king daily. He lived upstairs on the second floor of a large, gorgeously decorated room; the wooden pillars . . . supporting the ceiling were lacquered red. The walls were decorated with frescoes depicting Tibetan saints, and Tibetan rugs covered the floor. In the extreme end of the oblong room on a dais sat a life-sized, gilded statue representing Buddha, surrounded with burning butter lamps and vases with flowers and other bric-a-brac. There was, however, something phoney about that figure. The king noticed that I often looked curiously at the gilded figure . . . and pointing toward it said, 'That is my uncle. He died sixty years ago. I was given the formula for the preparation of such a body.'"

It would be easy to conjure a ludicrous portrait were it not that this overgrown child swathed in red and gold robes was a ruler of enormous power within his domain. Amused as he was by Chote Chaba's questions, Rock also recognized that he "was not a man to be trifled with," and told of a young man who made the mistake of taking to drink. The Muli king was a man of Puritan ethics who forbade wine, women, tobacco, and opium among the lamas. He once accidentally discovered that a sort of adopted son of his had become a heavy drinker and, in an outrage, sent for the offender. Upon receiving the message, the young lama replied, "If my head could carry me I would come; my feet cannot." It was a good try, but the king was beyond appreciating the humor of the response and ordered the man's head brought to him. The subjects, both lay and lama, lived in legitimate terror of him.

Muli was so rugged that it was impossible for the peasants to carry on any extensive agriculture except along the Litang River and in the narrow valleys of its tributaries. The tribespeople lived in shanty villages clinging to the mountains and, along with feeding themselves, supplied food to the king's household and the lamaseries. To a Westerner like Rock, the irritating aspect of the regional poverty was that it was entirely unnecessary because Muli, like Choni, was richly blessed with forests. "Any other country [but] China would certainly have exploited these forests," Rock complained, "rather than importing timber from the Pacific Northwest America while billions upon billions of feet of lumber go to waste. The natives fell these mighty monarchs of the forests, burn them, and use the resulting ashes as fertilizer for their buckwheat fields. Most of the timber is of hard yellow pine and dark spruce . . . Whole mountainsides are often burned over and no one cares."

The town of Muli, actually a lamasery with about 700 lamas, was presumably the capital of the territory but, in practice, the kings rotated their governments from Waerhdje (Wachin) to Kulu and Muli, the three largest lamaseries, which were all within a few days of one another. Kingdon Ward had presumed the ritual change of seat to have grown out of boredom but, as Rock observed, it was a practical necessity because the land at each place did not furnish enough to feed the king and his huge retinue in proper style for more than a year at a time. The peasants' situation in Muli was, then, somewhat different than in Choni: in one case the prince preyed upon the peasants to pay the Chinese; in the other, the lamaseries were the chief parasites. In Muli, Rock said, "the poor peasants stand before the ordinary priest officials with folded hands and bowed heads and answer only 'laso laso tidji,' a most deferential way of saying yes." Effectively, however, the results were the same, and the peasants of both Muli and Choni led ugly lives. It is no wonder that many of them turned to robbery or to selling their children into slavery.

Chote Chaba had a better grip on his domain than Prince Yang. The Chinese, having endowed him with ornate titles, left him in relative peace; there was no single tribe either large or strong enough to make trouble for him; he weilded enormous religious influence. Though he worried about being only a minor incarnation, unknown in Lhasa, the peasants collected and sold his feces in the belief that they were disease-preventing. He was, in Rock's words, "the most absolute ruler on God's earth."

The king was a strict disciplinarian, who kept his subject under his thumb by brutally enforcing pesky regulations. It was, for example, a crime for a man to be absent from his home for more than three days at a time. Peasants were not allowed to eat rice, wear trousers, or in any other way elevate themselves. Any person who had been born in Muli territory or had lived there more than one year was forbidden to leave it—this, according to Rock's informant, a gossipy lama of Yungning, "so that they would remain ignorant and regard the Muli king as the One Great Personage under heaven." Hence the regulation about absence from home for more than three consecutive days, which imposed considerable obstacles to flight from Muli. Muli appeared very medieval in character, and Rock, like other observers of similar situations in China, offhandedly applied the word feudal to its description. Feudalism, however, rested upon an entirely different system of land ownership; in Muli the peasants often owned their own land. The traveler's inclination originated in the sensation of having been cast back in time rather than in an attempt at historical exactitude. Land or no land, a peasant in Muli in the 20th century was no better off than a serf in Europe in the 11th century. If anything, in fact, the Muli peasant fared worse because the Muli king's rule was more arbitrary and one-sided than that of all but the worst of manorial lords. Chote Chaba bound the peasants to their land with so-called laws, permitted the lamaseries to bleed them, and collected taxes twice a year for his personal coffers; he gave absolutely nothing in return. The situation, Rock mused, was ripe for bolshevism and, since he viewed this as the ultimate evil, he "warned the Muli king of bolshevik agents and said should they ever come into his territory they would make enless trouble for him."

When Rock told him how the bolsheviks had murdered the Russian czar and his family, Chote Chaba expressed appropriate horror. He knew nothing about the Soviet revolution or even of Feng Yu-hsiang. He had, however, heard that in 1927 twenty Russians had arrived in Lhasa—despite a treaty between Tibet and China which excluded all foreigners except the British from Tibet—and offered to protect Tibet against the British and the Chinese. During their stay the Russians had purchased huge supplies of wheat and barley which they stashed in a large house. Rock explained that the Russians undoubtedly had designs on India and hoped to use Tibet as a stepping stone to the Indian subcontinent. The argument was lost on Chote Chaba. He knew even less about India than he did about bolsheviks, his idea of foreign affairs being relations with Yungning, the bandits of the Konkaling mountains, and Chinese who intruded into Muli to work the streams for gold. The story of the czar, on the other hand, convinced him that communists could not be trusted. He assured Rock that all was well in his domain. He said he would even like to resign his *t'ussu*-ship but that his peasants begged him to continue because there had been no fighting during his rule. "Perhaps," Rock thought, "he is aware of the trend of the times and would rather give up the reigns of government now . . . than be deposed by the Chinese." The explorer listened to claims of domestic contentment with skepticism; he saw signs that Chote Chaba lacked the conviction of his words. He considered it dangerous to give arms into the hands of the peasants or the lamas and shrewdly kept the army down to a couple of hundred men. Meanwhile, he stockpiled guns and ammunition with a collector's passion and showed a keen interest in a bullet-proof vest.

Laws in Muli were generally obeyed because it was as dangerous to be suspected of a crime as to be caught in the act. When a misdemeanor was committed anywhere within the territory, the names of all the suspects were brought to the king. Each name was inscribed on a separate piece of paper which was then rolled into a small wad and placed before the king. He withdrew one of the papers, read the name, and sent for that person to be thrown into a dungeon or to be beaten. There was no appeal; neither begging nor logic could save the subject from his sentence. Beatings were brutal. The victim lay flat on his stomach in a vestibule of the king's residence. The peasants who administered the blows raised his skirts—all Muli subjects wore skirts, and trousers or underwear were prohibited, perhaps to facilitate beatings—and proceeded to whip him with thirty paddle-like planks. Each plank was supposed to break on the second stroke; failure to break the paddle resulted in a beating for the beater. The victim usually fainted after the first few strokes. By the end of his punishment he had been mashed into pulp and bones and was flung aside for his relatives or left to die.

In dungeons, however, one did not achieve the luxury of unconsciousness. During all his years in China, no sight shocked Rock quite so deeply as that of Muli's prisons, and his decription of them seems worth repeating if only to prove that no nation or race has a monopoly on barbarism. "It is impossible to describe what I saw," he wrote years later. "Before me on the stone floor sat three creatures cross-legged: two Lolo tribespeople—a father and a son—and a Hsifan lama. Their heads protruded from huge square boards which must have weighed at least 75 pounds. The boards rested on the ground behind which, with stretched necks, squatted the inmates. They had been confined there for more than two years. Their faces could not be seen as long hair hung completely over them, so the lama [who guided me] picked up a stick and pushed their hair to both sides. What an apparition! Their glassy eyes had sunk deeply into their sockets, their faces were haggard, their heads no more than . . . skulls. Their grimy hands held the edge of the board which they shifted occasionally. It was impossible for them to lie down; they could not even scratch their heads or touch their faces. There they squatted in their own excrement, with rats and vermin as their companions. . . . When I inquired how they were being fed the lama explained that twice daily the jailor descended into the dungeon with balls of roasted barley flour mixed with water and a bowl of water. This had been their daily fare for over two years." A fourth prisoner, a Nakhi, was chained by the neck to the wall and, the lama guide assured Rock, was entirely innocent of the crime for which he suffered, his name having been placed in the king's lottery for having lodged a suspected bandit. The Lolos, however, had murdered a Chinese.

Horrified by his visit, Rock complained politely to the king who, having once denied the presence of any dungeons in his territory, was displeased to have been caught in a falsehood. When Rock said he wanted to photograph the prisoners, Chote Chaba erupted with laughter and observed that Rock would lose face by photographing such scum. He granted permission with much merriment and Rock, by this ploy, had the men released from the boards and moved out into fresh air for at least a few minutes. Eventually he talked Chote Chaba into freeing the Nakhi and removing the boards from the other three prisoners. He had not, however, made a dent in the Muli penal system; cruelty, like obesity, ran in the Muli royal family. The king, amused and perplexed by his concern, had only conceded to oblige the peculiar Westerner. Rock's faith in the king's word can be measured by the fact that he returned to the dungeon to make certain the promises had been honored.

GOLD IN THE HILLS OF MULI

If discipline was Chote Chaba's most persistent internal concern, gold caused his external one. The treatment which the citizens accorded the precious metal could not fail to dumbfound the Occidental mind. Where Westerners would have arrived in droves and carved great gashes into the mountainsides, the denizens of this hinterland scratched holes two or three feet deep into the soil or washed the gold from the streams in crude wicker trays and boards. The miners or, rather, scratchers, had to pay the king one tenth of an ounce of gold annually for this privilege and, though he grew rich on the gold tax, he carefully limited the number of prospectors. Stranger yet was that near Kulu, which was supposed to be particularly rich in gold, the king was reported to have had all the outcroppings covered with rocks.

Though gold nuggets were visible to the naked eye on the ground, the king's object was to keep Muli's gold resources as secret from the outside world as possible and to prevent the Chinese from descending on his domain. He was accordingly wary of all strangers who entered his territory; he had, in fact, originally wondered if Rock had not come to wash gold. Rock gave him and the lama officials $20 US gold pieces and told them there was plenty of gold in America, thereby calming his fears.

Two factors prevented Muli from becoming the scene of a Chinese Gold Rush. The territory's topography insulated it against the world, blocking major news leaks and discouraging

---

* Snow, in *Red China Today*, notes substantial ore deposits in this vicinty but makes no mention of gold. Hu, in *China*, indicates the presence of gold in western Szechwan but acknowledges that overall "China is poor in precious metals." The current regime has evidently tapped the hydroelectric potential of the mountains; I could find no recent data regarding the use of forest products.

prospectors. Secondly, Tibetan bandits roamed the mountains and always seemed to know when gold was being washed in one of the rivers; they would stay under cover, sometimes for as long as a year, until the washers finished and then descend upon them and make off with their gold. Chote Chaba claimed he was powerless to punish the bandits but, in fact, their activities were very much to his advantage. Chinese who heard about the gold usually heard about the Konkaling Tibetan bandits at the same time and thought carefully before venturing into Muli. The king had an unspoken understanding with the bandits: they did not attack him or the lamaseries while he did not attack them when they robbed gold washers. It happened that while Rock was in Muli during the summer of 1928, an ex-mandarin arrived in Yungning with about a hundred soldiers. He sent a messenger to the Muli king asking for permission to wash for gold, and a $300 loan, which he promised would be repaid by the fruits of the prospecting; also that the king notify the bandit chief not to harm the party. Chote Chaba replied amiably enough that the party could do all the gold washing it liked, that he had no money to lend, and that the mandarin himself should send a letter to the bandits requesting protection. To the Westerner versed in bandit lore, the latter suggestion seems outrageous, rather like alerting Frank and Jesse James to a large shipment of loot and then saying, "but please don't steal it!"

The old mandarin, however, was a sporting man, blessed with optimism. Since no Chinese could travel safely by himself in Tibetan territory, the mandarin hired a Tibetan to deliver his message to the reigning bandit chief, one Trashi-dzong-pon. Meanwhile, the Muli king had sent his own runner advising Trashi that the gold washers would be sending a messenger and to deal with him as he saw fit. The Konkaling mob, so called because their headquarters were near the Konkaling lamasery, therefore knew what to expect.

One would imagine that the bandits would have welcomed the washers, let them go about their business, and robbed them in the end. But the spectacle of a Tibetan carrying letters for a Chinese so infuriated them that they submitted the poor messenger—the only innocent party in the story—to horrible tortures and cut off one of his ears for him to take back to the mandarin as a response. Rock, himself en route to Kulu from the Konka mountains, found the messenger moaning beside the trail, dressed only in blood-soaked pants. He washed and bandaged the wounds and transported him back to Kulu, where the Muli king listened to the tale with transparent mirth and, only to appease Rock, offered the messenger a pair of shoes.

"Seems to me to be all a carefully laid plan to have this poor devil disfigured," Rock mused. "The Chinese will lose face badly if they let this go unnoticed, yet they certainly cannot come and attack these people for they certainly would be the losers. It is said that the letters [of the mandarin] threatened that if Trashi-dzong-pon would not provide protection, they would come with 4,000 soldiers and attack. . . . This, to my mind, is an utter impossibility, as the country is too mountainous, terrifically high, the rivers too deep, the trails narrow and next to unnegotiable, the passes high, the bridges few and far between. . . . The Tibetans, who know every inch of ground, have the advantage over the soldiers, for they can block them almost anywhere at will . . . The country is a natural fortress . . ." The Muli king, less confident of his domain's impregnability, fretted until it was clear that the Chinese would not seek to avenge their honor.

The king's dealings with bandit tribes sometimes took a baroque turn. Rock cited one incident when the king agreed they could traverse his territory unmolested in order to rob Yungning to the south. Then, inexplicably afflicted with either guilt or an attack of good neighborliness, he secretly informed the chief of Yunging of the pending raid and advised the latter to dispatch a hundred soldiers into Muli territory to fight the bandits. The full irony of the situation can only be appreciated when one realizes that the Muli king could have thwarted the bandits by the simple act of withdrawing the central part of a cantilever bridge on the only viable route to Yungning. The Tsungkuan of Yungning, wise to Chote Chaba's perfidious ways, would not risk sending soldiers into Muli; he did, however, take precautions to defend the lamasery against the marauders.

Given the king's imaginative diplomacy, he had little to fear from either the bandits or his peasants. He had reached a *modus vivendi* with the former; as for the latter, they were poor, terrified of the lamas, and deliberately unarmed. But, like all the so-called "barbarians" of western China, Chote Chaba lived in terror of the Chinese. Though they did not bother him with taxes and tributes, he worried day and night, and the stories Rock related about Feng's incursions into Choni only increased his anxiety. And indeed, in 1934, he came to what one charitably calls an untimely end after a series of bizarre events.

Chinese encroachment upon Muli happened almost accidentally, not, as the king had always feared, because of the gold. In 1929 Hu Jo-yu, the Szechwan warlord whom Lung Yun had chased from Yunnanfu in 1927 and who had been reassigned as Szechwan-Yunnan Border Commissioner (see Chap. III), espied an opportunity to grab the Yunnan governorship back from Lung. Lung succeeded in repelling Hu's forces, pushed them into the western part of the province, and stranded them on the north bank of the Yangtze in the Yungning region. Hu's troops, by then starving and bedraggled, ravaged the countryside for food and clothing, both of which were in short supply. As Lung closed in, the only avenue of escape from Hu was through Muli, and Hu applied to Chote Chaba for permission to cross his territory. It was at this point that the king's penchant for double-crossing proved to be his undoing.

He answered Hu's request in a friendly manner, sending several dozen felts for the soldiers' bedding and clothing, and stated, as his price, that Hu would have to give machine guns and ammunition in advance of his arrival. Needless to say, the king already had promised Lung Yun to trap Hu. Hu smelled a rat and tried to bargain to send the weapons later, but Chote Chaba insisted and, having no options, Hu complied. He marched in to Muli, where he was attacked by Muli soldiers in a narrow gorge—precisely the spot Rock had so often thought could be held by a handful of men. But the king had not made a clever bargain. Though Hu's forces were tired, they greatly outnumbered the Muli soldiers and overcame their territorial disadvantage. They took the gorge with great losses and then fought their way across Muli, harassed on all sides by snipers. Hu himself escaped on a raft in the Yalung river after the King had ordered the rope bridge cut. The fighting over and the Chinese gone, Muli returned to its quaint ways.

But some grudges die hard, and Hu never forgave the Muli king. Almost five years later, in September, 1934, Chote Chaba received an official letter advising him that Liu Wen-hui—the Szechwan governor, a former warlord supported by Chiang Kai-shek, had decided to raise his rank from "commissioner of the Barbarians," to "ambassador of the Barbarians" and that a delegation would arrive to perform the appropriate ceremonies on the 10th of the month. The king, always the overgrown child, had a weakness for flowery titles and orders; he had no inkling that Liu was a bosom buddy of Hu and that the conferring of titles was really a dark plot.

On the appointed day, the Muli Living Buddha cast the dice to read the king's horoscope. Like Caesar on the Ides of March, the king disregarded the inauspicious omens and proceeded to an alpine meadow three miles from Kulu where such ceremonies were traditionally performed. He took his entourage of lama officials, his nephew (the heir to the Muli *t'ussu*-ship), and servants; the Living Buddha, a firm believer in horoscopes, flatly refused to go along. Liu's official emmissary met the Muli party smilingly in the meadow, and the king served a feast in his tent. After the meal, when the presentations should have been made, the Chinese official began accusing Chote Chaba of being an enemy of the Chinese. Soldiers entered the tent and tried to seize the king and the lamas. One of the soldiers grew panicky and started a struggle; a gun went off and, in the ensuing melee, Chote Chaba wrenched himself free and escaped through the rear of the tent. The Chinese recaptured him and shot him through the head.

Rock received the news in a letter. Cruel, vain, and naive though the king had been, he had been good to Rock and had made the National Geographic expedition possible. The explorer felt terrible and observed that "somehow, I have always seen the end of these kings, buddhas, and princes, and visited them just in time."[4] Later he heard the story again from the nephew, the new Muli king, who had been on the spot. He, along with the lama officials, had been kidnapped and held for ransom for several months, but as the communists approached during the Long March in 1935, the kidnappers grew frightened and fled, leaving their captives to fend for themselves. The nephew made his way to a friend of Lung Yun's. After the "Red scare was over," as Rock put it, Lung escorted the nephew back to Muli and saw him installed as *t'ussu*. Lung, however, did nothing selflessly; his price was the incorporation of Muli under the Yunnan jurisdiction, formerly it had been in Szechwan. The new king sent gifts of golden buddhas and rugs, which Lung refused; he demanded that the king come to Yunnanfu in person to arrange the terms of his rule. The young king had no choice and duly appeared, stopping to see Rock, who was then also in Yunnanfu. Lung ordered a monthly payment of thirty pounds of gold. Two lamas who had accompanied the king went on to Nanking to plead Muli's case before Chiang Kai-shek. Rock never learned the outcome of the negotiations but noted that letters no longer bore the postmark "Muli, Yunnan" and suspected that Lung, for once, had been overruled. In any event, Muli was no longer the "impregnable" kingdom; the old order had gone, and the Chinese had a foothold.

THE BORDER LAND OF YUNGNING

Rock was a royalist at heart, but clearly on the side of benevolent royalty. The antics of the Choni and Muli *t'ussus*, fascinating and rich in gore as they were, displeased him, and he was quite convinced that both gentlemen had met just ends. Yungning, the third "semi-independent" territory of his acquaintance, impressed him favorably by contrast. It is "about the best ruled spot in Yunnan as it is outside the hectic Chinese territory. The Tsungkuan rules kindly but firmly. There is no bolshevism. The peasants on his approach prostrate themselves with folded hands and three times touch the ground with their forheads . . . No bolshevik agents have apparently influenced the peasants who are happy with their lot as they are protected from robbers and are ruled by people who are interested in their welfare—which cannot be said of people living under a Chinese magistrate."

This was an overstatement in some respects, and Rock eventually came to restrain his enthusiasm. He did, however, prefer Yungning to either of the other enclaves along the Tibetan borderland and he spent considerable time there between 1923 and 1933.

Yungning territory lay to the north of the Yangtze loop in northwestern Yunnan, along the trail between Likiang and Muli, the latter being its unpredictable northern neighbor. While protected by mountains to the north and the river to the south, Yungning was by no means a natural fortress, yet this particular district had eluded the attentions of the Chinese. The Yungning *t'ussu* was theoretically subject to the prefectural city of Yungpei to the south; at tax time this relationship was something more than theoretical. Otherwise, however, the Chinese were not interested in Yungning and left home rule in the hands of the local *t'ussus* or their designates. Yungning foreign policy, such as it was, centered about prolonging this disinterest.

Yungning was a wild country until Kublai Khan arrived in 1253 to establish Mongol order among the tribesmen. The territory originally comprised the Muli district as well as the sub-districts of Chienso and Tsoso. Muli and its rivers of gold went, as mentioned previously, as a gift to the hard-working lama who had converted the tribesmen from Black Shamanism to Yellow Sect Buddhism. Chienso and Tsoso, which had been farmed out as sub-districts to deserving relatives of the Yungning ruling family, were lost in the 18th century when the relatives proved to be less worthy than had been supposed. In 1710 they journeyed to Peking and paid tribute to the Emperor K'ang-hsi and were thereby appointed as chiefs independent of Yungning.[5] The remaining territory, that of Rock's day, was exceedingly mountainous, with peaks as high as 15,000 feet. There were a few arable valleys and, most significant in terms of agriculture, a high plain which supported some 300 families. This plain, Rock noted, could easily support three times that population if rice were grown and all the land used.

Such abundance, however, was precisely what the Yungning chiefs sought to avoid. The abbot of the lamasery once explained this negative strategy to Rock: "He said, 'we could grow beautiful rice, but if we did it would attract the Chinese and soon there would be no t'ussu or chief,' and the Chinese would take their land away from them. As it is they say 'Oh, it's too cold, no rice can be grown here,' etc. When the officials come to investigate the place, they buy them off, and [the officials] return saying 'Oh, Yungning is a terrible place. It's cold and it snows a lot and only little can be grown there.' Thus they are able to hold on to their land." Therefore maize, barley, buckwheat, and round peas were grown on the plain while rice was only cultivated furtively in one long valley. The peasants cured oxhides and deerskins to make shoes and tall Tibetan boots. Yungning policies being in some respects consistent, what applied to agriculture also applied to the villages: "They allow no improvements such as paved streets or erecting of shops, as soon the Chinese would come and they would own the place."[6] Nothing dramatizes the tribespeople's sentiments toward the Chinese so clearly as this deliberate self-denial.

Most of the inhabitants of Yungning belonged to the Hlikhin tribe, a branch of the Nakhi or Moso, and various subsects. Rock accordingly studied their history and culture as it related to his work on the Likiang Nakhis. Yungning territory, in fact, occupied nearly one-fourth of his scholarly two volumes, *The Ancient Na-khi Kingdom of Southwest China.* One notes with some surprise that Rock did not confine his racial generalizations to the pages of his diary. Even in academic texts he made startling statements such as "All Zher-khin people [a Hlikhin sect] are expert swimmers," an observation that competes favorably with 'All Negroes have natural rhythm.' It interested him that the Hlikhin and the Likiang Nakhis had grown so far apart that one could not understand the other's dialect and they had to communicate in Chinese. The Hlikhin had no written language. The Zherkhin, however, had a pictographic script which, though similar to that of the Likiang Nakhis was not understood by their *dtombas.* Yungning population also included a smattering of Lolos, Lisus, Hsifans, Chinese, and a floating number of pure Tibetans.

Like their counterparts throughout the western mountains, Yungning's peasants lived in extreme poverty, intensified by an eagerness to discourage the Chinese. Rock described a Lisu village as a "miserable conglomeration of huts, which look as if they had been dumped on a rock pile. The fields are poor and full of stones. . . . Most of the people of this village are afflicted with goitre, and cretins are common, as the 30 families constantly intermarry . . . The children go entirely naked and are black from filth." Some of the Chinese in the district were so poor that they sold their children into slavery. The Tsungkuan himself had bought a child from a Chinese family which could not support it; this was great luck for the child, whom the kindy Tsungkuan treated as one of his own.

Thanks to the original missionary lama, the prevailing religion in Yungning was Yellow Sect Buddhism. Spiritual life, however, was definitely on the decline. Lamas were supposed to be celibate but, in Rock's time, Yungning had the reputation of a place where a lustful lama could live in peace. There were among the lama population renegades who had fled Lhasa in moral disgrace. Rather than dwell in the lamaseries, the lamas only went there for festive occasions. They did not marry, but they did not refrain from sexual activities. There was, therefore, an inordinate number of bastard children, and the Hlikhin vocabulary did not include a word for father. The Tsungkuan himself had once been a lama but took himself a "wife" and enjoyed earthly pleasures. Following the religious example, the moral codes among the lay population were relaxed. "In Yungning it is the girl who remains in the home, and she takes unto herself a boy as husband whom she keeps as long as he works, and as long as she enjoys his presence. She can send him away at any time and take unto herself another husband." At this point Rock could not refrain from judgment: "The result of this promiscuous sexual intercourse is an enormous amount of syphilis and other venereal diseases. The moral standard of Yungning is thus anything but high." But, with scientific fairness, he noted that "suicide, so common among the Li-chiang Na-khi is, . . . totally unknown among the Yungning people." The social structure Rock observed in Yungning occurred quite frequently among Tibetans in western China; many tribes even practiced polyandry. Cressey suggested that polyandry evolved as one kind of arbitrary birth control in a geographic situation where food-producing land was minimal; similarly, the lama system, with each family sending at least one son to the lamasery to take vows of chastity, theoretically restricted population growth.[7]

The ruling clan of Yungning claimed proudly to be of Mongol origin, directly descended from the officers left by Kublai Khan to govern Yungning. The family bore the name of A which, according to legend, it had acquired quaintly in the fifteenth century when one of its members went to the court of the Emperor Yung-lo. "As he was kneeling before the Imperial presence he was asked his name; not understanding the question, he replied, with folded hands, 'A,' which in the Hli-khin language is as much as to say 'I do not understand.' Thereupon the Emperor conferred on him the family name A."[8] During Rock's time the *t'ussu* was one A Ying-jui who preferred the opium pipe to power. He died in 1927, in his son's minority. The governing authority was his relative, A Yun-shan, the man who invariably appeared in the pages of Rock's diaries as the "good old Tsungkuan." As a youth, A Yung-shan had taken vows of celibacy and studied for six years at the Gaden lamasery near Lhasa, but when he returned to Yungning to take on its administration he gave up his religious life. He was altogether a likable man and Rock mused that, had Fate placed him in the West, he might have made a successful diplomat, for he was high in native intelligence and political skills. Fate, of course, had done nothing of the sort, and even the Tsungkuan was afflicted by ignorance and supersitition, artistic concoctions of fact and fantasy to explain the world. On a practical level, his lack of formal education made the business of government cumbersome. Neither he nor any of his officials could read a single Chinese character. In a crisis, "letters are brought with charcoal and chicken feathers attached, which means great importance and that runners go day and night. The letters are brought to the island in the lake and there given to the Tsungkuan, who cannot read them. He has to wait until the next morning until the wind has died down, and he is then rowed ashore. And by the time he gets to Yungning the greatest part of the day is gone. He has to hunt for someone who can read letters and then for someone who can write a letter—which is an entirely different proposition . . ."

Yet the Tsungkuan's manner of ruling moved Rock deeply. There were neither courts nor jails, and people were permitted to do as they liked, apart, of course, from growing rice or improving the villages. An individual with a complaint came in person before the Tsungkuan and stated his case from a kneeling position. The old man listened with fatherly patience. Rock witnessed a number of these audiences and always marveled at them. The problems varied widely: people told of their children having been kidnapped by Lolos from Muli; a man complained that his wife had run away with another man; there were disputes over property and money. The Tsungkuan's justice was most effective in situations where he could decide a quarrel between two parties, and he arbitrated with the wisdom and tact of Solomon. His gentle humanism, however, could not save children from their Lolo abductors, and he did not have the military forces to send a raiding party to Muli to fetch them. In this case he would notify the Muli king and hope for some cooperation.

A Yun-shan exercised his imagination in dealings with the Chinese. After he had spent some time in Yungning, Rock realized that the Chinese officials were not always fooled by the rice that had not been planted or the ramshackle villages. But, inasmuch as they received squeeze money to report on inhospitable conditions, they found it lucrative to overlook the truth. The Tsungkuan, who was no fool, played at this game expertly and often managed to outmaneuver his adversaries. A Chinese official who visited Yungning while Rock was there "*sub rosam* had asked one of his . . . men . . . to mention casually during the conversation [with the Tsungkuan] . . . that Yungning was a large place . . . and that undoubtedly this would be the place where Chinese officials [should] reside and that such a recommendation should be made to the powers-that-be. This would be the only reason that the Tsungkuan would be scared and give the official a big bribe to say that this place is too cold and produces nothing, etc., and to avoid placing the official here. But he did not reckon with the Tsungkuan, the latter a very wily individual and old. Apparently he does not care any more, for he knows his days are numbered. [He said,] 'An excellent idea. I am old and cannot protect the peasants any more. You call your Chinese official. It would be the best thing for the country.'"

The bluff paid off this time. The Chinese official went away without his customary pay-off but unwilling to jeopardize the possibility of future squeeze. Rock, harboring illusions that Western diplomacy was conducted in a straightforward fashion, criticized the whole affair; he had always thought the Tsungkuan an honest man and could not permit himself to approve such deceitfulness. Asian politics were all the same, he concluded mournfully: "both sides are conscious that neither believes the other, but this is all done with perfectly straight mein." If the Westerner did not appreciate the Tsungkuan's cunning, the neighboring *t'ussus* did, and he was often called upon to mediate in delicate situations.

Unfortunately, much of what the Tsungkuan had said to the Chinese official was true and, indeed, he could not protect the peasants. He had no army to speak of, and raiding parties such as the Konkaling bandits often descended upon Yungning to meet little resistance. Nor could he mount any defense against the tax-collectors. They demanded payments years in advance, and he was forced to pay. This, too, was a game, the Chinese with their prediliction for bombast asking wild sums, and the Tsungkuan satisfying them with, perhaps, a hundredth of the amount. He absolutely refused, however, to plant opium in his territory, though he smoked an occasional pipe himself, and this defiance cost him. The Chinese sent an official to estimate the quantity of opium that could be grown in Yungning and taxed him according to their figures for what he was not growing. In 1931 he simply ignored an outright order from Lung Yun to turn the fields over to the poppy.

A Yun-shan was old and weary of years of arbitrating and bargaining, bandits and taxes. He recognized that the good old days of the independent *t'ussus* were drawing to a close and confided sadly in Rock that he had made preparations to flee as soon as the Chinese showed signs of moving in. Unlike the rulers of Choni and Muli, he did not run about frantically playing both ends against the middle, trying to preserve his waning authority; he observed with dignified resignation. Power no longer attracted him, and he sought tranquility in his personal life. An earthy old fellow, he stashed all his money, precious belongings, and "wife" on the island of Nyorophu in the middle of Lake Kulu along the Muli-Yungning border and, whenever his presence was not indispensable in Yungning, happily repaired to the island. This was in every respect a practical arrangement. A Yun-shan loved his pretty, sweet-tempered young woman and their small children and reveled in domestic serenity. Moreover, the island was precisely in the center of the lake, out of rifle range from the shore and, thereby, a safe refuge. Nyorophu was very beautiful, and Rock—who was also keen on personal safety—camped there pleasantly on several occasions when he considered the neighboring countryside dangerous. The Tsungkuan's habit of withdrawing to his happy family resulted in a certain inattention to affairs of state, but a spectator such as Rock could not help but feel that he deserved at least this reward for the services he had rendered his people.

Rock had no confidence in any of the other important personages in Yungning. The *t'ussu* was an opium sot; the Tsungkuan's secretary, besides being illiterate, was an avid student of bolshevism; the highest ranking Living Buddha was as "intellectual as a cow" and believed glass could be manufactured from water; the abbot of the lamasery, from whom Rock essayed to understand the mysteries of local religious ceremonies, lacked concentration. "He counts the beads of his rosary continuously," Rock complained after a frustrating day. "He carries on a conversation while he prays and counts his beads. Either his head is at praying and not in conversing or the opposite, but still he talks, and every silent minute. Or while I talk to him [he] . . . is praying." The Tsungkuan, therefore, was left to administer single-handedly.

By the 1920s and '30s, the administration of Yungning amounted to fending off predators, and the Tsungkuan had grown pessimistic about his ability to keep up the fight. Yungning's foes became stronger and more numerous while the Tsungkuan's resources for combatting them correspondingly dwindled. With such a friendly neighbor as the Muli king, who permitted the Konkaling bandits to pass freely through his territory en route to Yungning, one did not need enemies. The *t'ussu*-ships of Chienso and Tsoso were still in the hands of the A family, but complicated household rivalries prevented either district from aiding beleaguered Yungning. Provincial authorities only cared about what they could wring from the district in taxes and squeeze.

Confronted with depressing odds, the Tsungkuan persisted earnestly in his effort to protect his people first against the Chinese, by his peculiar diplomacy, and second against invasions of bandits and militarists by bribes or force. In 1924 he ordered a tamped earth wall to be built around the lamasery in Yungning against the Konkaling bandits and their periodic visitations.

When warned in advance of their intentions, as by a letter from Muli, he posted his few soldiers in strategic places and armed the peasants to protect the lamasery. He did not, one remarks, fear popular revolt and had no qualms about distributing weapons among his people. In this fashion he sometimes averted a sacking, but one hamlet after another was laid to waste, either by bandits or equally rapacious Lolos. Robbers seldom satisfied themselves with looting. It was customary for them to burn villages on their retreat and to kill any soul with the temerity to protest their advances. A year or more might pass, the peasants would have rebuilt their poor houses and replenished them as best they could; then the bandits would descend upon them again. "On February 9th, 1929," Rock related, "the eve of Chinese New Year, when the poor Hli-khin peasants had decorated the gates of their homes with pine trees, the savage Lolos came and robbed them, drove off their cattle, and set their homes in fire; worse still, their children were carried off into slavery. And this happened less than 10 miles from where Chinese New Year was celebrated in great style in Yungning. Passing by . . . on my way to Muli in March of the same year, a few of the peasants were sitting in the ruins of their former homes; silently, they contemplated their former belongings."

"This state of affairs is of course mainly due to lack of communications, the wildness of the country, and the long distance, comparatively speaking, from the governing center of these outposts of China. But if the truth must be told, it is also due to the indifference of the Chinese officials."[9]

When it was not Lolo or Konkaling bandits, it was someone else. The Tsungkuan, like the Muli king, had to deal with Hu Jo-yu's renegade army in 1929, and though he managed to do so with less bloodshed than his sometimes friend and neighbor, it was costly for him. Lung Yun, naturally, had alerted the Tsungkuan that he was under no conditions to assist the rebel. Hu, meanwhile, had taken Yungpei, to which Yungning was responsible, and started sending charcoal and feather-adorned letters to the Tsungkuan. He had exhausted the food supply in Yungpei and demanded that Yungning send him rice, clothes for his soldiers, and money. The Tsungkuan, fearing that Hu's army would descend upon his district, sent a token of several mule loads of rice. Hu was not pleased and issued a new letter requesting 30,000 silver dollars; if the Tsungkuan did not oblige him immediately, he threatened to burn Yungning to the ground and kill every living creature. The demand was outrageous, "like getting money from a piece of wood. . . . The poor Tsungkuan, knowing Chinese ways of bombastics, had to send some money so he gathered together 400 dollars in nickels, which amounted to about 110 dollars silver, to which [Rock] added some old Yunnan paper banknotes worth a couple of dollars. To me it seemed strange, but the Tsungkuan was certain that nothing more would be heard about the 30,000 silver dollars, and in this he proved correct." Hu's army traversed Yungning on the run for Muli and did miraculously little damage.

After Hu had been defeated, Lung Lun ordered one of his sons as military officer to Yungpei; when the latter learned that the Tsungkuan had provided even miserly assistance for Hu, he had an excuse for squeezing Yungning even further. He commanded the presence of the Tsungkuan who, suspecting foul play, pleaded old age and sent the abbot of the lamasery (he of the incessant mumbling and praying) in his place. Lung's son arrested the lama and held him for several months until the Tsungkuan, at his wits' end, relented and paid heavily in gold for his release.

A Yung-shan was the best friend Rock ever had among the people of China. With him the Westerner felt nearly on an equal footing; he often said that the Tsungkuan had been a father to him. When the old man died of natural causes during the summer of 1933, Rock sensed a personal loss and, in 1942, during the war with Japan, he went back to Nyorophu to stand at his graveside, the location of which was not known to the Tsungkuan's two youngest sons. Rock had once predicted that, after the Tsungkuan, Yungning would be overrun by Lolo bandits. Instead, though the A family continued to hold the traditional offices, Chinese influence became strong, and Lung Yun's henchmen made inroads into the district. Rock cast his gaze over the fields of opium poppies and understood the change.

VIII

TO MINYA KONKA FOR THE
NATIONAL GEOGRAPHIC

Rock was in Yunnanfu in the spring of 1927 when he learned about the death of Professor Sargent; the information was contained in a curt note from Ernest Wilson, who had temporarily assumed the administration of the Arnold Arboretum. Badly informed and worried about the institution's financial future, Wilson determined to discontinue all excessive expenses, of which, he believed, Rock was the most prominent. He ordered the explorer to return immediately. This development disappointed Rock mightily. Bandits, soldiers, civil war, insults, promises never to return to China, and suicidal thoughts forgotten, he could not bear the prospect of the Western world, and he had counted on talking Sargent into extending his field work. Harvard thus eliminated as a potential sponsor, Rock turned his charm on his old acquaintances at the National Geographic Society, to whom he proposed an expedition in the vicinity of Muli.

He needed, first, to do a little mending of fences. He had been dissatisfied with the Society's editing of his article on Muli of which he had received copies in Choni, and had discharged fiery complaints to Ralph Graves, the magazine's assistant editor, who rightly and reasonably resented the strong language and the accusations of incompetence.[1] Rock belatedly regretted his haste and apologized; his nerves had been on edge, he explained. "Don't be offended on account of my last letters. My bark is worse than my bite, and I am not so bad as the letter might have you think I could be . . . Now let's forget bygones and let's be friends." As a token of amity he enclosed a photograph of himself dressed in the Choni prince's Tibetan costume. "Unless I can work in the wilderness and the unexplored regions," he concluded, "I . . . have no incentive to living, and I can now understand why poor Meyer committed suicide."[*]

The powers at the National Geographic had had sufficient experience with temperamental outbursts to dismiss the suicide hint as Rockian histrionics, and they were attracted by his proposal to explore unmapped mountains in and around Muli. Rock wanted to stay in China and was already ordering supplies and equipment before any firm agreement had been reached. He promised to study the tribespeople, take photographs, collect plants and zoological specimens, and prepare maps. The officers of the Society felt that, particularly with such a mercurial character as Rock, matters should not be handled by mail. Albert Bumstead, the Society's cartographer, was not enthusiastic about the quality of Rock's Amne Machin maps, which were now at the Society, and wanted to give him a few technical pointers. Several other details, including finances, merited careful discussion. Rock was requested to return to Washington to make final arrangements. He grumbled, but had no choice. He left Yunnanfu in mid-July with his Nakhi photographer and taxidermist in tow —at the Society's expense.

The visit went badly. Rock had little time to readjust to "civilized" society and persisted in playing the potentate, supported, on this occasion, by the servile attendance of the Nakhis. They, it appeared to observers, were in Washington less for the purpose of learning taxidermic and photographic techniques, as had been suggested, than for following their master about. The Society's president, Oliver La Gorce, sensed that Rock was not in a cooperative mood and discovered that he did not respond to

---

[*] Frank Meyer, a Dutch-born plant collector for the USDA was drowned in the Yangtze in June, 1918. The cause of his death was never ascertained, and suicide was only a theory.

subtleties. They had a disagreeable telephone conversation. "I . . . [pointed] out," La Gorce reported, "that it was the fundamental policy of this Society and its officials to do everything in its power to uphold the hands of the man in the field and to give him every human cooperation, and that he, on his side, must realize that he was not transacting business with coolies in the interior of China, and that there must be a sound understanding between individuals in order to get the best possible results on all sides."[2] Rock thereafter became a bit more flexible, at least in the company of reigning officials. Bumstead, lower on the Society's totem pole, had less success. "In my conversation with Dr. Rock," the cartographer told La Gorce, "I suggested some of the simplest ways of finding latitude and longitude in the field. His reaction to these suggestions was not encouraging. He 'did not know much about mathematics,' 'All those things would be too much bother and take too much time' and 'perhaps he should not have tried to make a map anyway.' But after a little encouragement, he said he 'might like to take a course in that sort of observing.' Personally, I doubt if a person of his temperament would make a success of it."[3]

But for all Rock's difficult personality and unprofessional approach to exploring, except for plant hunting, which he did in a business-like fashion, the Society valued Rock's past record and talents. Graves summed up their feelings neatly when he said, "Rock is one of the world's finest photographers and is a resourceful explorer and geographer. At the same time, he is one of the most cantankerous of human beings."[4] Advice from the State Department to the effect that Rock would run a risk by going back into the Chinese interior was ignored.[5] By November 20 Rock had returned to Yunnanfu as a National Geographic explorer. It was too late in the season to head for the mountains, and he prepared to wait out the winter in the provincial capital.

Rock spent a restful four months in a rented house, socializing and gossiping with the foreign elements of the city and observing the wily Lung Yun in action. People who knew of his plans to proceed west to Likiang and beyond told him he was mad. Bandit activity on the caravan trail between Yunnanfu and Tali had reached an unprecedented peak, and Lung's power was not established in the hinterlands. All but a few obstinate foreign missionaries had evacuated the mountains during the troubled spring. Rock toyed with the idea of going by airplane as far as Likiang, but did not dare send his men and supplies alone. He decided, however, to ask William I. Hagen, a young lawyer and Vice Consul at the American Consulate to join the expedition. He told Graves: "He is really a good boy, . . . clean and enthusiastic, he likes the out-of-doors, having been brought up [on] a South Dakota farm."[6] Rock could be persuasive, and Hagen responded with enthusiasm, resigned his consular post, and signed on for a "period of apprenticeship," at the rate of $100 a month. Rock justified the deal to the National Geographic Society, explaining that he could train Hagen to carry on his work, perhaps to lead exploring groups in the future.

The expedition finally left Yunnanfu on March 22, a month later than Rock had planned, owing to unsettled conditions along the road. A huge farewell party assembled to watch the departure. The whole foreign colony turned out, even the Japanese. The caravan mules wore red tassels and bells; along with Hagen, a Tali missionary named Père Eschartz, thirteen Nakhis, and the muletiers was a military escort of thirty. Everyone was in a festive mood.

The caravan encountered nothing more fierce than some wild rumors on the road to Likiang, but the uninterrupted suspense made the trek "all but pleasant."[7] Six stages from Yunnanfu, at the town of Chuhsiung, Rock and Hagen stayed two nights in the Presbyterian mission run by Miss Cornelia Morgan, one of *the* Morgans, a rare missionary who made a favorable impression on Rock. She was, at the time, about 45, all skin and bones, and almost blind; her English was so interspersed with Chinese idioms as to be almost unintelligible; and she combined toughness and kindliness in a winning manner. Chuhsiung, Rock noted in a comic aside, was one of those Chinese places possessing a telephone, but one could not call to the next town though there was a phone there, too, because the second instrument did not have a bell and no one had bothered to install one. The only way of making contact was if someone in the second town happened to lift up the receiver to see if anyone was calling. It was, Rock observed, a classic example of the Chinese *pu yao ching*, "never mind."[8] In Tali Rock developed digestive difficulties and prepared himself for death. Hagen made himself useful running errands while Rock recovered in the hospital. A month and a day after its departure from Yunnanfu, the party made a triumphant entry into the village of Nguluko. The Nakhis embraced their relatives while Rock, who never had anyone to embrace, looked on enviously.

The National Geographic expedition lasted around two years. Between April and September, 1928, Rock and his party left Likiang-Nguluko and went via Yungning to Muli; during the summer they made two excursions to the Konka mountains northwest of Muli. In September they returned again via Yungning to Likiang. During the winter of 1928–29, from November to February, they went back into Muli by the same route and returned to Likiang once more; no proper exploring was undertaken due to heavy snows. They left Likiang for the third time in late March of 1929, through Yungning to Muli and thence northeast to Kangting; returning south via Minya Konka, they reached Kulu in mid-July and Yungning by early August. For the remainder of 1929 Rock concentrated on studying the Nakhi tribe.*

The expedition would never have been possible without the cooperation of the Muli king who, in exchange for guns, ammunition, and the novelty of occidental company, asserted his authority with the Tibetan bandits. Rock's party, therefore, was left free to roam in Muli territory and to take the road to Kangting —the robbers' favorite hunting grounds. But, as Rock pointed out, to Grosvenor, "this is not a picnic to Mt. Sinai."[9] Chote Chaba dispatched runners to the bandit chief, Trashi-dzong-pon, asking him to see that Rock received safe conduct. But the king,

---

* The names Rock used may be confusing. What he calls the Konkaling or Konka range are mountains between the Chochi Ho and the Wuliang Ho, between Muli and Patang, Long. c. 100°10′, Lat. 28°30′. The highest mountain in China proper, the famous Minya Konka or Kungka shan, is northeast of Muli, east of the Yalung river, and near Kangting, which used to be called Tatsienlu. It is impossible to find many of the place names that Rock used on any contemporary map. Since the areas he explored were often blanks on his own maps, he assigned names according to information he received from local inhabitants and thus often used a Tibetan name. When he knew there was also a Chinese name, he gave both. Consult his maps in the *National Geographic Magazine*, LX: 1931 and LVII: 1930.

not quite trusting Trashi's word, sent his brother-in-law, also chief of the little Muli army, to accompany Rock on his first visit to the Konka range in June, 1928. Chang Tui-chang, the brother-in-law, was not at all helpful. He was so scared of the bandits that he tried to rush Rock for a quick glimpse of the mountains and then dash back to Muli. Rock, of course, had other ideas but had to compromise because Chang's whining made him nervous. They spent more time than suited poor Chang, but less than satisfied Rock. The explorer returned, therefore, in August, leaving Chang behind.

Trashi and his gang, it turned out, were in the course of their annual circumambulation of the holy mountain range, like any good Tibetans; Rock and his gang were exploring in the opposite direction. Rock met the terrible Trashi head on, had a cup of tea with him, and received a gift of yak butter and cheese. The bandit leader behaved amiably, promised not to bother Rock, and indicated himself as temporarily occupied with religious conventions. Rock relaxed and took his time. Later, when an unseasonal hailstorm hit the mountains and destroyed the barley crop, the Tibetans, who had heard that Rock's party had gone around the sacred peaks in the wrong, or cursing, direction, blamed him.

When Rock returned to Likiang in September he sent a coded cable to his sponsors in Washington to advise them of his progress. "Expedition discovered mighty snow range next Amne Machin highest in western China, height between 24,000—25,000 feet, called Konka Risonquemba, consists of three isolated magnificent peaks on high plateau. Mapped region and the Shouchou [Wuliang] river course hitherto unknown. Secured several hundred color plates besides other negatives, and thousands of zoological and botanical specimens from this previously unexplored and unknown region controlled by Hsian Cheng bandit chief."[10] The following letter to Graves was even more expansive: "It was a most wonderful experience, even more so than on the Amne Machin journey, for the mountain range was not known to exist . . . It will be a great surprise to geographers. . . . You must agree with me that it is a great feather in the cap of the Geographic to have been successful in such a difficult quest. Such mountains are not discovered every day."[11]

Indeed they are not, nor were they then. Kingdon Ward had already seen them and called them by name, "Gang-ka-ling," in a 1924 publication.[12] Like Rock, he had had a distant view of another great mountain, "almost due east, probably on the other side of the Yalung;" this was undoubtedly the great Minya Konka. Rock had once referred to Kingdon Ward's attempt to visit the Konkaling; by September he conveniently overlooked it.[12] "Discovery," then, was a bit off the mark, though contemporary maps were blank in the region and dotted lines indicated the presumed courses of rivers, so Rock indeed performed a very valuable geographic service. Unfortunately, however, by overstating his accomplishments, he reduced them. The same thing happened with the matter of altitudes. Rock's propensity for hyperbole and his refusal to use precision instruments led him astray. He restated his miscalculation of the Amne Machin and compounded his mistake by making a bad guess on the Konkaling peaks; not one of them exceeded 21,000 feet. Rock transmitted his findings with the enormous self-confidence of the dedicated amateur and came closer to the spirit than the letter of the question.*

Inflationary prose or not, the first season's exploration had produced impressive loot: 243 color plates, 503 black & whites, thousands of plants, and over 700 bird skins. Rock was accordingly satisfied with himself. Had it not been for Hagen, he would have had no complaints, but the young man was beginning to get on his nerves. "For example," he explained, "I let him go with the caravan the last few days and told him to be sure to stay behind the last mule to see that nothing is lost. The day before last of arrival at our village, he rode ahead and told one of the men to stay behind. The man, of course, did not stay behind, and the caravan arrived at the camp minus one load of trunks containing

---

* Rock did not publish his Konkaling story until 1931 (*National Geographic Magazine* LXL: 1–65), and his article was far more modest than his field reports. He brought down the heights of the mountains to around 20,000 feet (p. 47) and acknowledged earlier travelers in their vicinity (p. 64). There is no way to account for the discrepancy in altitudes between the 1928 cable and the later article because Rock had not revisited the range. One can only guess that he readjusted his figures after having seen Minya Konka.

color plates and negatives. The mule carrying the load drifted behind and was taken by Lolos with the intention of stealing it, but they became afraid and tied the mule to a tree in a side valley, where it was later found. How easily such carelessness can result in the utter failure of an expedition. . . . I have never lost a single thing on any of my expeditions . . . [Hagen] could never lead an expedition, and he expressed the sentiment that he would not care to lead one as it is too nerve-wracking and too strenuous work to be on edge all the time." Rock's catalogue of Hagen's shortcomings ran on for pages; there seemed to be only two things in his favor, that he was essentially likeable and could write fiction, a talent for which Rock had little respect. Though he did not mention it to Graves, Rock was unhappy with Hagen's political attitudes. "I am tired of listening to 'everybody is as good as everybody else'" he wrote in his diary. "This is usually put forward by people who have attained nothing, have no experience, know little, and possess still less. . . . [Hagen] would like to make a bolshevik out of every beggar." So far, however, Hagen still had his uses and had not sinned sufficiently to cause Rock to discharge him.

Since winter snows eliminated the prospect of exploring or collecting, Rock decided to pass the season in Muli and Yungning to see what more he could learn on the cultural side and to take photographs. He also found time while in Yungning to write a long article on the Konka range and its colorful bandits. Hagen volunteered for the dirty work and stayed behind for a time to make sure the collections got through to Yunnanfu, whence they could be shipped safely to the United States. Rock was enjoying the hospitality of the "good old Tsungkuan" in the third week of November when the first real misfortune struck his caravan. On the 21st one of the muletiers suddenly became ill and died within a few hours. Rock, who had never lost a man before, was terribly upset. He guessed correctly that the muletier had died of relapsing fever, a disease transmitted by fleas and ticks. The men, muletiers and Nakhis, were all frightened, and Rock did not have any medicine to treat the fever, which he knew to be highly contagious. Later the same day, a second muletier had a subnormal temperature, one of the symptoms, and Rock hastened to enforce sanitary measures among the men. He ordered anyone who handled the corpse to wash thoroughly with disinfectant; those who did not obey he discharged without ceremony. The ailing man eventually died on his way home. Otherwise the winter passed without excitement, and Rock accomplished a good deal of work.

THE ROOSEVELT EXPEDITION IN YUNGNING

One day at the end of January when he was comfortably ensconced in the Tsungkuan's residence on the island of Nyorophu, Rock heard some curious news: a caravan from Likiang reported that "the Indians with long beard had arrived in Likiang with half-dead mules, many of which were scattered along the road abandoned by the muletiers."[13] Rock thought this bit of information over for a while but, not knowing what to make of it, put it aside. A few days later a messenger was rowed across the lake bearing a pencil scrawled note for him, telling of the arrival of a foreign exploring party in Yungning. The note was signed by Col. Theodore Roosevelt, Jr. and his brother, Kermit. Rock, who had met Kermit once in New York, received the message with delight and replied that he would meet them on the trail.

The Roosevelt caravan arrived in Yungning on February 3 and stayed for three days. The men had come to China mainly to hunt the giant panda, which makes its home in western Szechwan. The party, financed by William Kelley and backed by the Field Museum of Natural History in Chicago, included C. S. Cutting, a photographer, and Jack Young, a Hawaiian born, American-educated Chinese, who served as interpreter and guide.[14] Herbert Stevens, a British plant and animal collector, had left the caravan at Likiang and taken his own route. The Roosevelt party also had retained a peculiar crowd of Kashmiri, Tibetan, and Chinese hunters and helpers.

Rock, starved for news of the outside world, feasted on the Roosevelts' talk. They told him, among other things, that Hoover had been elected U.S. President; he knew nothing about it. But catching up on current events did not prevent him from noticing that the brothers looked shabby—bearded, muddy, and tattered. Rock recognized the disadvantages at which unkempt appearances placed the Westerners with the local inhabitants, who

KANCHOW (CHANGYEH) TIBETAN CITY IN KANSO PROVINCE.
Arnold Arboretum of Harvard University

A STREET IN KANCHOW *Arnold Arboretum of Harvard University*

A MARKET IN LIKIANG. *National Geographic Society*

A STREET IN LIKIANG. *National Geographic Society*

CHINESE SHEPHERDS IN THE GRASSLANDS.
Arnold Arboretum of Harvard University

NOMADS ALONG ROCK'S ROUTE TO THE KOKONOR MOUNTAINS.
Arnold Arboretum of Harvard University

NAHKI SWIMMERS WITH
GOATSKIN RAFT.
National Geographic Society

A LOLO CULPRIT
IMPRISONED IN MULI.
National Geographic Society

A SONGMA OR MEDIUM FROM YUNGNING. *National Geographic Society*

NAKHI WOMAN IN
TRADITIONAL COSTUME.
National Geographic Society

HLIKHIN WOMEN IN
CEREMONIAL ATTIRE.
National Geographic Society

IDOL IN BUTTER, FROM BUTTER FESTIVAL IN CHONI.
National Geographic Society

NAKHI ASSISTANTS
CHANGING DRIERS ON
PLANTS AT ROCK'S HOUSE
IN NGOLUKO.
National Geographic Society

ABBOT OF YUNGNING
LAMASERY AT PRAYER
WHEEL HOUSE.
National Geographic Society

found it difficult to believe that these shaggy specimens were indeed President Roosevelt's sons, as their documents stated. That young Theodore had been Assistant Secretary of the Navy did not impress them either because, as Rock remarked, the mountain people had absolutely no idea what a navy, much less an ocean, looked like. Rock was flabberghasted when he heard the Roosevelts had not stocked food provisions but had, more or less, lived off the land. This, Rock knew from experience, besides being rather unappetizing, might be dangerous as they proceeded north to areas where food was scarce. It also, no doubt, explained the sickly condition of the caravan animals; instead of exchanging mules along the way, the Roosevelts had come from Bhamo with the same animals and had not brought enough paddy rice (rice in the husk) to feed them. Furthermore, "they had no sleeping cots, no folding chairs, no folding tables, bedding but no bed sheets, and their pillows were as black and shiny as if they had been polished by a bootblack."[15] The level of disorder and discomfort appalled the veteran Rock and brought out the squeamish in him, but he found their lack of regard for local etiquette even more disarming. When they told him they planned to pass through Muli and had brought some Woolworth medals for the Muli king, he blanched and warned them to give the king a wide berth as such trinkets would insult him. They did avoid a face-to-face meeting, but the Muli king knew of their presence in his territory and sent them a riding horse and a sturdy mule as gifts; when the Roosevelts returned to the United States they responded with an Army rifle and ammunition on Rock's advice.[16]

Rock thought Cutting "a queer sort" and disapproved of the way he worked—"snapping here and there without a tripod" —but he got along famously with Kermit.[17] He helped the group get fresh mules, arranged for their passage through Muli, advised on northern routes and, despite all his criticisms, felt bad when they left. He loaned them two Nakhis to help them in Muli.

Indirectly, however, the Roosevelt expedition caused aggravations for Rock. A week after it left Yungning to go north, Rock headed south to Likiang, where he had left Hagen to look after the collections. "On my arrival in Nguluko I found all O.K.

and sent Hagen to Likiang to fetch supplies and mail which had arrived. While I was alone in our small house, I looked about among the papers lying on the shelf, wishing to write a letter. Among the writing sheets I found an open letter written by Hagen to the American Consul at Shanghai, and I saw my name mentioned on the page. Naturally, I read [the letter] . . ."[18] (Naturally? Rock's record of snooping into other people's mail was all but clean. Reporting Hagen's case to Graves, he felt it necessary to explain how he happened to read the letter.)

Rock thus learned that Hagen had met Stevens, the Roosevelts' naturalist, and had become friendly with him. Stevens had offered to hire him, and Hagen had accepted. The letter announced his intentions to sever his connections with Rock and join the Englishman. If there were any malicious references to Rock, the latter did not mention them. When Hagen returned from town, Rock was in a rage and summoned Stevens. Hagen, he said, "turned white and most disagreeable and was really vicious"; Stevens "was of course very mad, too." But Rock was the only one of the three who knew how to turn a temper to his advantage and so intimidated the others that their partnership never materialized. He made Hagen leave for the coast the next day with the collections, as a final task. He also considered it necessary to inform the Roosevelts of the incident, the implication being that Stevens had tried to steal his assistant. Then, to make sure that Hagen would be *persona non grata*, he wrote the whole story to Graves. He swore for the umpteenth time never again to travel with a white man.

A second unpleasant by-product of the Roosevelt visit did not surface until autumn, when Rock heard second-hand that Jack Young, the Roosevelts' interpreter, had published an article in an English-language newspaper in Shanghai in which he intimated that Rock had involved himself in politics along the Tibetan borderland. Such privileges as Rock's roaming freely about Muli were cited as evidence of subversive activities. Rock was alarmed when he learned about the article because he always had been careful to steer clear of politics and now feared he would run into problems with Chinese authorities. Fortunately, the *China Journal* had picked up the article and countered it by stating that they were "surprised that an American paper is lend-ing itself to such suspicions. Dr. Rock is and has been engaged in thoroughly straightforward scientific exploration work. . . . Such rumors have been spread about every explorer that ever traveled in China, and we might just as well ask whether there was not some ulterior motive to the Roosevelt brothers' expedition . . ."

Rock, naturally, had no kind words for Young or his prose —"a piece of impudence from a Bolshevik youth," he said. He informed the Roosevelts, who knew nothing about it, and the National Geographic; they were all sympathetic. Nothing came of Young's article, and Rock's subsequent difficulties were in no way out of the ordinary. "Still," as he remarked to Graves, "it hurts."[19]

FIRST REPORT ON MINYA KONKA

In March, 1929, when things were beginning to thaw, the Rock expedition, now minus Hagen, left Likiang Nguluko for the third time, again heading north to Yungning and Muli. Rock had seen Minya Konka from the Konka range and wanted a closer look. He planned to proceed from Muli to Kangting. The ever affable Muli king once more interceded on his behalf, and the Konkaling bandits were requested to leave him in peace. The robbers had not forgotten the matter of the hailstones and the barley crop and, had Rock shown his face near their mountains, they undoubtedly would have murdered him on sight. But they did not object to his traveling to Kangting, and it was profitable business to stay on Chote Chaba's good side. The road was tranquil, and Rock found time for light reading. He had taken *David Copperfield* along, and Dickens evoked grim childhood memories.

The collecting, especially, of birds, was excellent, and Rock worked efficiently. But the crowning event was his visit to Minya Konka, which he mapped, measured, and photographed to his heart's content. The mountain had been known to the Chinese for centuries. The first Westerner believed to have traveled in its vicinity was the Catholic priest who placed it on the map he prepared for the Emperor Kang-hsi's imperial atlas.[20] Count Bela Szechenyi, a Hungarian explorer, saw it from a distance in 1879 and guessed its altitude at 7,600 meters—about 10 meters over the actual height. The China Inland Mission map of China published in 1899 shows an unnamed peak, obviously Minya Konka, at an altitude of 24,900, an overestimation of only nine feet; yet, for some unfathomable reason, the C.I.M.'s 1923 map does not indicate any mountain there at all. James H. Edgar, an Australian, drew sketches of the peak from Yingkwanchai, about twenty-five miles to the northwest, and in 1923 reported optimistically that "nothing but a scientific measurement will make me give up the 30,000-foot hope."[21] Rock once again harbored hopes for something higher than Everest because his own sighting from the Konka range had been spectacular; he did not, however, approach Minya Konka with any undue expectations.

Rock was at Minya Konka during the second week of May and again in mid-June, 1929. He made gorgeous pictures of the mountain and neighboring peaks—one of which he diplomatically named Mt. Grosvenor—and went about his geographical business. Using a boiling point thermometer, aneroid barometer, and clinometer, and a prismatic compass, he took bearings and estimated the altitude of Minya Konka from three separate observation points at elevations of 18,600, 18,500, and 17,200 feet. However unsound his methods of ascertaining altitudes, nowhere in his diary did he indicate that he thought Minya Konka higher than Everest, nor did his subsequent letters to the National Geographic Society contain any such suggestions. Yet suddenly, on February 27, 1930, he composed a cable stating unequivocally: "MINYA KONKA HIGHEST PEAK ON GLOBE 30250 FEET ROCK."[22] He sent the message from Haiphong as he was on his way back to Washington, and one can only surmise that—provided he trusted his own figures—he had held back the information for fear of its being intercepted in China. For an explorer to discover the world's highest mountain would have been rather like finding a gem larger than the Hope Diamond, and some precautions might understandably be observed in disclosing the coup.

One can conceive of the excitement at National Geographic headquarters when Rock's wire arrived, but it was not a foolhardy organization. Without studying Rock's notes, the Society was unwilling to commit itself to a statement; specialists such as

Bumstead were certainly skeptical. Rock soon arrived on the premises, and the matter was submitted for immediate discussion. He made initial concessions and admitted the possibility of error due to the necessarily crude methods he had used to establish the mountain base line; he had measured it by time of travel (horse gait). Officers of the Society looked over his calculations and concluded that "from the data he was able to obtain, his observations point to the possibility that the height of Minya Konka may rival or even surpass that of Mount Everest, . . . Dr. Rock makes no specific claim for the preeminence in point of height of Minya Konka, but he presents to the scientific world the results of his observations with the precautionary statement that calculations of the height in actual feet will be subject to revision by other expeditions with more elaborate surveying equipment."[23] This still rather positive statement became the subject of further deliberation, no doubt prompted by A. Heim's accounts of Minya Konka, which he visited shortly after Rock. By the time Rock published his findings, in October, 1930, the mountain had shrunk by almost a mile to 25,600 feet—still a bit over the mark but much closer to reality.*

Returning south from Minya Konka in 1929, Rock again called on the Muli ruler in his residence at Kulu. Before he left —as it turned out, forever—Chote Chaba loaded him down with gifts: banners, buddhas, a rug, golden bowls, not to mention four ounces of gold dust. Rock responded with the king's favorite toys, rifles and ammunition, and proceeded toward Yungning. The king's lama-secretary accompanied him to the Muli border and, at the moment of parting, burst into tears. He caught Rock in a bad mood; the Westerners was stone cold to his tears which he dismissed as products of a "wily Oriental" eager for gifts. Lamas in general, he decided, were a "greedy rotten lot"; their morals and their bodies "both stink."[24] The secretary received nothing for his display.

---

* Minya Konka became a painful subject for Rock. After the Sikong Expedition an American party of Terris Moore, Richard Burdsall, and Arthur Emmons III made the first climb of the peak in 1932. Moore, who lived in Cambridge, Mass., heard that Rock was in the neighborhood and went to the Arnold Arboretum to visit him. When he brought up the subject of Minya Konka, Rock instantly switched horses and plainly did not want to discuss the mountain.

Only when Rock reached Yungning in the second week of August did he learn that Hu Jo-yu's renegade army had arrived in the neighborhood and that the Yangtze ferries had been smashed, thereby cutting him off from Likiang. He reacted with customary panic and imagined himself kidnapped for ransom. The Muli king, alerted to his dilemma, sent him a message to return to Muli and wait there until things calmed down. Rock considered the suggestion but was always slow to make decisions under duress. This time his hesitation served him well because it soon became known that Hu intended to take his forces into Muli territory, and the king withdrew his invitation. Rock repaired grimly to the island of Nyorophu as the Tsungkuan's guest, watched events unfold, and made cynical observations about Chinese politics. "Chiang Kai-shek should have waited a little bit in his celebration of the unity of China. . . . Oh America, when will you awake and learn the truth? Instead America is fed with silly propaganda of the great things the Nationalists are accomplishing—and this mainly by the missionaries who ought to know better but fear that if they would actually tell the truth the contributions [would] stop coming."[25]

Rock also produced feverish reports for his sponsors in Washington, describing the local conditions down to the last lurid detail. The officers of the Society, receiving his bulletins several weeks after mailing (by which time, of course, Rock had arrived safe and sound in Likiang), had trouble judging the extent of his plight because he had often cried wolf in the past; besides, they could do nothing to help him. However, as a minimum precaution, they forwarded one of his letters to the State Department where it was read with curiosity by China experts. Stanley Hornbeck, in his acknowledgment, noted casually that Rock had extricated himself from difficult situations in the past and would undoubtedly do so again.[26]

By the end of September Rock was comparing himself to Dreyfus on Devil's Island and Napoleon on St. Helena; Nyorophu had become a trap for him. Hu's army drew closer and closer to Yungning, the demands upon the Tsungkuan increased, and the latter grew more worried each day. It occurred to him that Rock might make a good intermediary if matters reached a crisis, and he dropped a few hints along those lines. Much as he sympathized with his host, Rock wanted absolutely no part of Chinese politics. The fear of having some unwelcome role thrust upon him finally dislodged him from the island. If the Tsung-kuan did not find some means for him to cross the Yangtze, he threatened, he would go north into Konkaling country, try to come down through Chungtien, and there search for another way over the river west of Likiang. "To tell the truth," he admitted to Graves, "if we had decided to go that way it would have been as much as if we had all committed suicide on the spot," which he did not intend to do.[27] The bluff, however, fooled the Tsung-kuan—who, perhaps, never expected any devious rhetoric from an Occidental—and he arranged a crossing.

Rock and his caravan assembled on October 13 on the north bank of the Yangtze. "It took us two days to cross; twenty-two men worked like slaves from early morning until after sunset. Stark naked with only an inflated goatskin tied to their stomachs, they would plunge into the stream and push us across the river sitting on a couple of inflated goatskins. Three would pull and two would push, and yet we were taken downstream for over a mile; coming in contact with a whirlpool, we would spin around like a merry-go-round. It was exciting, I can assure you." One mule, which carried nothing of value, was washed away in the current; everything else—men, animals, trunks, photographs, collections—crossed safely. Likiang soldiers, alerted to Rock's approach, met him on the other side of the river and, with appropriate pomp, escorted him back to Likiang.

The exploring phase of the expedition had ended, and Rock tallied the results. He had seeds of 317 species of plants, among them, 163 of *Rhododendron,* and over 30,000 herbarium specimens, which he had shipped to the Department of Agriculture; the bird specimens numbered 1703; he had taken nearly 900 autochrome pictures and another 1800 in black and white. The specimens and photographs were all of first quality, and Rock took great care in packing them for shipment to the United States. Moreover, he promised Graves he could write enough to fill at least one entire issue of the *National Geographic Magazine.* He did not, he said, feel like sharing the magazine with a gondola trip through Venice or a tour of old Poland. In fact, he was even contemplating a book of his adventures. Graves countered by citing the Society's obligations to a popular audience, but he mentioned that financial support might be available for Rock to prepare a separate scientific account of the expedition. Technical writing, Graves suggested, was much more Rock's line. His popular articles were an editorial headache and had caused measurable grief among those whose business it was to render them into a publishable form. One harried editor, William Simpich, complained bitterly to La Gorce about Rock's manuscripts:

"Imagination he has none. Nor form. By lack of form I mean he has never yet submitted a finished article; that is, an article sustained by any well-thought-out structure, having a comprehensive lead, developed arguments, and a climax or rattle to its tail.

"He has never yet submitted a job of writing which consistently followed any central theme. He can do more, with surplus words, to confuse readers than any contributor whose work has ever come to my attention anywhere.

"Apparently he has never learned to write with a view to holding reader interest to the end; there is no unity or singleness of impression in any of his work. He does not know how to convince readers that he has something to say . . ."

"Added to this basic defect, his sentences often include six or eight clauses and without expressing a complete thought. Mr. Graves knows all this and will corroborate it. But for the toilsome re-write in this office, nothing of his could ever have been published. It would have made the *Magazine* look ridiculous.

"Of course, he does not believe this, and complains that we 'change his meaning.' If this ever occurred, it was because no one could tell *what* he meant." But, the same editor concluded with equal conviction, Rock's raw material was of enormous value and could be extracted from him in conversation. Then, if one prepared an outline for him, he returned with an almost presentable article which, with editorial readjustment, could be published.[28]

Graves spared Rock the humiliation of outright criticism and tried to steer him toward the style which suited him best; in this effort, as Rock's publications testify, he was absolutely justified. Rock, unfortunately for adventure lovers and his bank account, could make an exciting story dull and confusing with no effort at all. He saw everything and recorded with a scientist's precision, but he lacked the ability to establish priorities and it hurt him to omit anything from his narratives; the only order he understood was chronological. At the same time, he had seen much and, unlike many travelers, understood what he had seen; one could not help but encourage him to make his knowledge available to the world. Rock chose, ultimately, to do his writing in the scholarly framework natural to him. He never entirely gave up his idea of a popular book because people unfamiliar with his limitations kept urging him to write one, and visions of staggering royalties were always seductive. But eventually he grew discouraged with his rambling prose, referring to his aborted effort as his "funny book."

INCREASING HOSTILITY TO FOREIGNERS

During the autumn of 1929 in Likiang-Nguluko, Rock's interest in Nakhi culture really blossomed. He had promised the *National Geographic* ethnic novelties and decided to photograph one of the *dtomba* ceremonies for it; he selected a colorful routine for exorcising demons, complete with chanting, dancing, and the blood of freshly slaughtered chickens, and paid the sorcerers for their special performances. They, in turn, explained the symbols and chants as written in their ancient manuscripts, which he always called "books." Rock retained one of the sorcerers, a dim-witted type, to help him make synopses of the books and got through more than 500 in a few months. In the course of this work the idea of translating the ancient script fastened on his brain as a possibility for the future. He was nearly 46, and active exploring was beginning to take too much out of him. He thought about spending a year in America or Europe to write his book; then he hoped to talk Grosvenor into financing one more expedition. After that, provided he could raise money for it, he could live in Nguluko indefinitely and concentrate on the Nakhis. Planning was hard for him because he had not yet lined up financial support; and then, one could never count on China's political future.

In principle he liked living in Nguluko in his comfortable house, which he decorated with the spoils of his travels. He could see the snow mountains and the alpine meadows, and often rode out into the hills for relaxation. His cook and servants cared for him. He could work in peace and did not mind the people who came to him for medical help; he treated them kindly and quite expertly. There were a few missionaries in Likiang, and he was especially fond of a Miss Scharten, whom he visited often.

For a man as easily aroused as Rock, however, there was always some obstacle to sustained tranquility. One day, for example, he was in Likiang when a commotion began in the streets. Two young girls, it was said, had drowned themselves in the lake. Rock, curious, hastened to the scene with his Nakhi servant and there, indeed, were two pale figures floating Ophelia-like on the water. Since the native people in the assembled crowd were too superstitious and frightened to touch the bodies, Rock ordered his servant to fish them out with a pole and pull them ashore; he felt their pulses to make certain they were beyond medical help. He could do no more and returned home, more or less depressed by the afternoon's activities. The following day relatives of both girls arrived on his doorstep. He prepared himself to receive their thanks for his assistance and instead listened with astonishment as they accused him of having stolen bracelets from the dead girls' arms. He shouted at them and threw them out of the house in a fury. By evening he had reached a state of "nervous breakdown" again.[29]

Three weeks after the drowning incident, Rock quarreled with one of his servants over money and concluded miserably that "the more one does for them, the worse it is. . . . One cannot take them and treat them like a white man."[30] He took these arguments dead seriously and judged offenders on his own terms, always forgetting between unpleasant incidents that he was in a strange moral and ethical climate of his own choosing, always surprised and insulted when people did not behave according to his European expectations. A sense of humor, which would have spared him much agony, left him in a lurch.

By the end of 1929 Likiang seemed to Rock to have become unfriendly, and he found his excursions from the village into town increasingly disagreeable. The streets were now full of soldiers left over from Governor Lung's battle with the rebellious Hu. Having little else to do, they amused themselves by sneering at foreigners. Some of the Likiang youths, following the soldiers' example, also took to mouthing anti-foreign slogans as a sport. The atmosphere in Likiang, if not as frightening as it had been when Rock traveled through Szechwan in the spring of 1927, was hostile. Most of the missionaries, with guilt and God on their side, took the taunts philosophically; but Rock was touchy and upset. A rumor that extrality would be abolished in 1930 alarmed him. "One will be treated like a Chinaman," he predicted unhappily to Graves "and one will have to travel without escorts and be a certain prey to bandits without a recourse to complaint. I wanted to get out before this happened but the fighting and rebellion etc., [have] made this impossible."[31] The rumor, like so many others, proved false. While the Nanking government secured new treaties with smaller powers ending their extraterritorial privileges, France, Britain, Japan, and the United States maintained them; American citizens in China were thus protected until 1943.

The Chiang Kai-shek regime in Nanking did not impress Rock at all which was not surprising, considering his experiences of Lung Yun's provincial manipulations and of various intrigues along the Tibetan borderland. "A lovely republic indeed," he wrote Graves. "I wonder when America is going to wake up to the fact that there is no such thing as a republic in China. . . . Whenever I get back to America I will tell the people not to be fooled by those Kuomintang propagandists who paint such rosy conditions in China and the great strides which have been made. Republic? Nonsense, a bandit nest from one end to the other; and I for one shall be glad to get out of it. I have had enough."[32] Chiang, to him, was just another warlord, no better and no worse than any other. That he allied himself with Feng Yu-hsiang—whom Rock loathed—during the Northern Expeditions confirmed his evilness in Rock's eyes. That Chiang and Feng fell out and became enemies in 1929 only seemed like a tiresome repetition of the old warlord battles of the early 1920s. Rock had heard that hundreds of thousands were dying in Kansu while neither Feng nor the Kuomintang lifted a finger to end the slaughter. It was difficult for him to observe the new regime charitably, and it pained him to read Chiang's praises in the foreign press.

By the end of 1929 "getting out" seemed to Rock like a very good idea. He was tired of tensions and intrigues; his hair had turned gray and his stomach bothered him frequently. He spent a last emotional evening with his favorite Nakhi servants and the *dtomba*, whom he plied with wine; the next morning, January 13th, his birthday, he rode out of Likiang on the long road to the West. "I am convinced that I will come back once more," he had written in his diary,[33] "and I shall never leave this place until my remains shall be carried to be fed to the flames and my ashes to the winds."

ROCK OBSERVES THE NAHKIS, THE KUOMINTANG, AND THE MISSIONARIES

From the moment Rock arrived in San Francisco, on March 21, 1932, he felt out of sorts. "Life in these far away wilds makes one perhaps a little queer, and being thrown all at once into such hectic surroundings as America has to offer, with a Washington heat thrown in for good measure, is apt to make one a little unhappy," he suggested belatedly in one of his more lucid moments.[1] He went directly from California to National Geographic headquarters in Washington, where the first item on the agenda was the question of Minya Konka; then there were sessions over manuscripts and maps. The society's officers, by now accustomed to his volatile properties, handled him gently and tried to lead him into producing matter appropriate for their magazine. But the most delicate hints of criticism upset him, and he could not admit graciously to any imperfections. Nor was he pleased with the lack of enthusiasm when he declared he could write articles *ad infinitum*. Graves patiently tried to explain that the magazine's popularity depended upon its universal appeal and that the Society could not afford to overdo a good thing. Rock took everything as a personal slight and retreated into a defensive shell.

His frame of mind did not improve upon hearing that Wilson at the Arnold Arboretum had scorned his botanical collections. "It seems that Wilson wants to get credit for everything," he remarked bitterly. "Personally, I do not care, . . . Wilson wants to keep his name before the public whenever possible."[2] Of course, he cared very much indeed, and Wilson's criticisms were not fair.[3]

Rock's main objective was to find support for his work in China, but he had arrived on the scene of the greatest financial crisis ever to beset the United States. Money was tight and there was little left for effete cultural studies. Though he had heard about the Crash of 1929, he had been unable to appreciate its impact. He had counted on the National Geographic Society to renew its contract with him and, in fatter years, it might have done so in spite of his difficult personality. When the officers observed that the Society was not interested in his translations of Nakhi religious scriptures, he volunteered an expedition into Lolo country. President LaGorce responded with an offer to keep him in the field for one year at a salary of $2,000 less than he had received for his most recent expedition. Rock was affronted and bargained for more, but LaGorce would not increase his figures. Rock argued aggressively and finally worked himself into a corner where he could not, on principle, accept reduced conditions. He approached the Department of Agriculture with botanical projects, but had little luck there; they, too, had little money to spare and were wise to his temperament. The Library of Congress, another potential sponsor, was interested in the Nakhi religious manuscripts he showed them and agreed to purchase more for their collection, but the financial benefits to Rock were negligible. Next he tried Harvard. The Arnold Arboretum would not take him on but, finally, the Museum of Comparative Zoology offered a small sum—a pittance, really—to keep a few Nakhis in the field collecting bird specimens. Otherwise, Rock had to fall back on his own resources which, though adequate, would never support him in his old age if he had to keep backing his own research. The negotiations, especially with the National Geographic Society, left a bad taste in his mouth for many months and brought to his mind the German proverb *'Der Mohr hat seine Schuldigkeit gethan, jetzt kann er gehen'*—'The Moor has done his duty; now he can go.' "It hurts me," he confessed to Graves, "that after all I have done I have been sort of, well, I don't need to say the word, . . ."[4]

Rock, who had expected to return to the United States as a conquering hero, suffered double damage to his ego. He made the usual round of dinner parties and gave a few lectures with great success, for he was as brilliant in conversation as he was dull in manuscript, and his listeners hung on his words. There was one really bright moment. In May he was invited to Baylor University in Texas to receive an honorary degree, which entitled him to use the precious Dr. he had affected for years. But much as he enjoyed attention it did not compensate for the lack of official appreciation, as expressed by the problems he had in raising financial support.

The heat and humidity in Washington became insufferable by June, and the evils of industrialized society weighed heavily upon him. Rock worked doggedly repairing his manuscripts and identifying plants for the Department of Agriculture while day-dreaming about mountain meadows above Likiang. He had fulfilled his literary obligations to the National Geographic and was desperate to get back to China. No further offers of assistance seemed to be forthcoming, so he decided to risk living off his savings, which he augmented a bit by selling some of his accumulated Chinese and Tibetan artifacts. He left the continental United States in June without a single pang of regret.

On his way back to China, Rock stopped briefly in Hawaii, staying as usual with his old friend Lyons. In the islands he was already a minor legend, and great excitement attended his presence in society; the competition among hostesses for his company was fierce. Though he had always frowned on social conventions, Rock found it agreeable to be celebrated, particularly after his difficult visit to the mainland. He patched up outstanding quarrels with the University of Hawaii and checked the progress of plants he had brought to the islands years before. His old patron, Alexander Hume Ford, inspired by his presence, gave birth to a new brainchild: a botanical garden on the island of Hawaii, to be set up on a strip of land running from the top of Mauna Kea to the Hamakua coast and thereby comprising several different climatic zones in a small area. Ford made an announcement to the press after Rock's departure, and the *Honolulu Advertiser* for July 4, 1930, ran a headline asserting Professor Rock Will Return to Direct Great Botanic Garden.

"The Pacific Institute of Botany and Acclimatization," the story continued, "one of the greatest botanical ventures in modern times, is to be for four years under the leadership of Dr. Joseph Rock . . . according to an announcement made by Alexander Hume Ford, director of the Pan Pacific Institute, who received assurance that he would return in 1933 from the distinguished scientist before the latter sailed from Honolulu for his home in Tibet." In spite of journalistic superlatives, Ford's project was loaded with 'ifs', and never got off the ground. Rock, meanwhile, was not at all committed to the idea of giving up China for an administrative post. But with the knowledge that one could always change one's mind, it was reassuring to have some options.

Rock lapsed into a peculiar state of inertia when he reached Yunnanfu in the fall of 1930. He behaved like a semi-invalid, lying in bed with a book in the morning—often Dickens novels—and strolling about the city in the afternoon. He purged himself with emetics, paid unnecessary visits to the French Hospital, and wrote out a will with the full expectation that it would shortly be of use. He had called ten of his Nakhis from Likiang to wait on him. He drank afternoon tea with Dr. Watson and the other foreigners and gossiped about local politics. At night he had disquieting dreams. He half-heartedly drafted more maps for the National Geographic, but he made no effort to resume his Nakhi studies. He ran out of enthusiasm and motivations. He wanted vaguely to go to Likiang, yet he did nothing about it. Political events being, if anything, more chaotic than usual, he had no trouble inventing excuses for his indecisiveness. And then his health was far from perfect. He had no serious diseases, just his old digestive problems, headaches, and pains in the hand he had once injured in a riding accident. But he worried about Likiang where he would be several days from any decent hospital. From October to February, for more than five months, he did practically no work of any sort. Though he could account for his inactivity to others, he could not explain to himself why he delayed his journey to the interior. He was dissatisfied and uneasy at best, suicidal at worst. In short, he had arrived at a crisis of middle life, and nothing pleased him, least of all himself.

The idea that he would never undertake another major exploring trip crept up on him like a poisonous vine and spread despair in his system. He brooded over his failure to excite American sponsors and felt unappreciated; in panicky moments he decided he had wasted his life. One day he wanted to go back to Hawaii and work over his Chinese notes; the next he thought it might be nice to retire to some quiet European library; the third day he yearned for Likiang. But, as though weighted with lead, he could not move himself in any direction from Yunnanfu. He had just fled Western civilization with contempt, and there were quite a few Westerners in the provincial capital; still, he craved intellectual companionship. Never had his sense of friendlessness or displacement been so acute, or his longing for a place to call home so keen.

Rock had never liked Yunnanfu very much; during 1930 he liked it less than before. Anti-foreign sentiment was much more conspicuous than it had been in Likiang, due to the concentration of Westerners in the capital. As if it were not bad enough to be unwanted by the local population, he also managed to alienate Harry Stevens, the American Consul, of whom Rock wrote in his diary: "He clings like a barnacle to the ship of state." Stevens apparently accused him of being a bad American because he sang foreign operas and spoke foreign languages with other nationals. Rock, no doubt, did not let this comment pass unchallenged.

Yunnanese politics provided the only diversion from Rock's misery. Lung Yun's armies spent a lot of time fighting anti-Kuomintang warlords in Kwangsi province to the southeast. That Lung's interests in subduing Kwangsi rebels were entirely self-serving was evident even to American observers. "The Yunnanese are theoretically supporting the Nanking Government and are supposed to be under the orders of the Cantonese General Chen Chi-tang," noted Nelson Johnson, Minister in China, to the Secretary of State. In practice the chairman of the province, General Lung Yun, seems to be acting more or less independently. Yunnan controls the province of Kweichow, and as both are relatively poor provinces they are fairly safe from invasion and have therefore no desire to invite one by participating more actively in the civil war. Yunnanese cooperation with Canton has therefore been very half-hearted, except at Nanning, provincial capital of Kwangsi, where they have done most of the fighting. Their expedition into Kwangsi has been in the nature of an 'opium war,' merely to open up a safe and convenient trade route to the sea for Yunnanese opium."[5]

Every time Lung's armies absented themselves from the home province, bandits had a field day and watchful rivals plotted to seize Yunnanfu, all of which made foreign residents feel insecure. There were also suspicions of communist activity. In early October, during one of Lung's southern adventures, Rock reported a "Red scare" in the city. "Soldiers from outlying districts have been rushed to the town and over 350 people have been arrested, said to be Bolsheviks. Fifty of these special Bolshevik agents from other provinces who carried on their persons instructions as to go about in the Bolshevik uprisings have been bayonetted to death in the most gruesome manner during the night . . . On them were found 75 photographs of the Chinese officials here and rich people, whom they were to assassinate; also a list of people who were in the game. Bombs also have been found in various quarters, and the town was to have been burned yesterday. However, so far nothing has happened, but everybody is very careful and sleeps lightly. No kerosene can be bought as its sale has been prohibited, and those who had some on hand had to turn it in."[6]

One cannot discount the possibility that the Yunnanfu agitators were not communists at all but henchmen of one of Lung's warlord adversaries. Since Chiang Kai-shek had declared war on communists, it was easy for his partisans to rid themselves of opponents by declaring them Bolsheviks and inventing evidence to back their charges. Rock, for example, while thoroughly disliking Chiang, had been sufficiently swayed by the opinions of foreign newspapers and missionaries to believe that anyone who did not support the Nationalist regime might be communist. He believed that Lung's missions in Kwangsi were strictly anti-Red, whereas the Yunnan governor was really Chiang's military errand boy, battling warlords with personal ambitions for his own gains. (Communists, trying to scatter Nationalist forces, did actually give some support to the Kwangsi rebels in Nanning, but never enough to justify the argument that the insurgents were communists). Rock, insensitive to subtle political hues, swallowed the story of the communists' attempted coup in Yunnanfu with absolute credulity.

During the troubles, the American Consul advised his charges of hiding places in case of emergency. Rock found the disorder unnerving. He would have been safer in Likiang, out of the political mainstream, but he stayed in Yunnanfu, rescued from his sluggish mood by proximity to danger. "Things change over night hre," he wrote Graves in November, 1930, "and it is perhaps these uncertainties which keep one keyed up and make the struggle of life worthwhile. I don't know. I have lived on excitement for the last ten years, and a humdrum existence is next to unbearable to me." Meanwhile, using the pretext of arranging an expedition into Lolo territory, which neither the National Geographic nor the Department of Agriculture had any intention of backing, Rock befriended Lung Yun. The Governor, himself a Lolo, promised escorts, safe passage, a special Lolo passport, and letters of introduction. The journey never materialized, but Rock amused himself by being received by Lung in the *yamen* and observing the man.

It was Edgar Snow, finally, who roused Rock from his stupor. The young journalist arrived in Yunnanfu at the end of December, 1930. Rock met him and conceived an instant dislike for him. Snow, he quipped scornfully in his diary, "is an uncouth American youth, inexperienced and learned in ill names . . . he carries the general air of the provincial American ignorant of his own ignorance." Snow described his proposed trek from Yunnanfu to Tali and then south into Burma; he planned to travel on foot with two missionaries and a couple of coolies and to live off the land. Rock laughed and predicted he would either be murdered by bandits on the way to Tali or die of food poisoning. Having found accidently a motive, "ignoramus" or not, for travel, Rock started to organize a caravan and told Snow he could join; Simpson and Hagen were forgotten. Snow, though impatient, took his advice and waited until the end of January. Until the day of departure, Rock wondered whether to go to Hawaii instead but he was, in the end, overjoyed to leave

Yunnanfu, where he had been offended by jeers every time he went out in the streets.

Snow, who had never traveled caravan style before, was dazzled by Rock's stately progress. "During the march," he recorded, "[Rock's] tribal retainers divided into a vanguard and a rear guard. The advance party, led by a cook, an assistant cook, and a butler, would spot a sheltered place with a good view, unfold the table and chairs on a leopard skin rug, and lay out clean linen cloth, china, silver and napkins. By the time we arrived our meal would be almost ready. At night it was several courses ending with tea and liqueurs." When Rock became ill with a fever and was too weak to ride, he hired Szechwanese chair coolies—the Yunnanese would never stoop to carrying a chair—thus adding to Snow's embarrassment. The military escorts and Rock's private arsenal, however, made sense to Snow when he realized that the pack mules carried $10,000 in silver and other desirables. They met caravans that had been robbed, and bandits were reported everywhere by the villagers.

Rock's disposition, which had been bad in Yunnanfu, deteriorated along the road, and Snow's company did not help matters. "Never again shall I ask anyone to go with me," he reminded himself. Snow's manifest sympathy with villagers and peasants grated on him tiresomely. His experiences with anti-foreign name-calling were fresh on his mind and he had lost all patience with the Chinese. His indictments of filth, lice, and indifference became violent, and at night he filled his diaries with abuse. "People, white people, are depraved enough and silly enough to want to save the souls of such a despicable people. They have souls no more than lousy, flea-ridden dogs have. Why not let them sleep in their filth and ignorance and leave them alone, let them rot in their bandit-ridden country? The best thing that could happen to the world would be a huge catastrophe which would annihilate that miserable, mean, selfish, filthy, degenerate race . . . What a fool I was to come back and be insulted by such vermin. If I shouldn't go back to civilization this year, somebody ought to kill me."

Snow's wide-eyed compassion made Rock a little ashamed of his own intemperence, and he saved strong language for the pages of his diaries, but the effort of restraining himself in public made him boil over with hatred in private. Paradoxically, in one of their conversations Rock told Snow, "Of course, there'll be a revolution in this country. It's bound to come some day and it will be the bloodiest in history. I don't want to be around then but I won't blame them. Look at that lad with us we're taking back to Likiang! He was only 14 but he was impressed into the army to replace another fellow whose mother could bribe the officers. They led him off with a rope around his neck and gave him a gun and now he's back from [Kwangsi] with that crippled arm. Had to crawl back to Yunnanfu by himself, and when he came to me his arm was full of maggots! . . . How long will people put up with that kind of business?"[7]

Snow, a man of patience and perception, recognized in Rock an eccentric genius of unpredictable temper who, from long periods in isolation from Western civilization, had tried to immunize himself against the miseries to which he was exposed. For all his complaints, Rock laughed out loud at the bawdy songs of the Szechwanese chair-bearers as they picked their way through mud and stones; he tended the sick and paid his escorts their opium money and extra so they would not steal from the villagers. The frequency and intensity of his rages against the Chinese suggested several levels of frustration: that the people endured incredible abuse without fighting back; that, short of his amateur medical ministrations, he could do nothing to help them, and that, whatever he did, they would never accept him in their society. Vivid recollections of his own humble beginnings generated within him a rather uncomfortable sense of identification with the Chinese downtrodden. He loathed the Chinese as he loathed all that was sordid in his past; at the same time he championed the principle of revolt against sordidness. Snow, though ignorant of Rock's family history, spent enough time with him to realize that he was a complicated man and suffered his proclamations with humor and intelligence.

Rock had good luck again along the trail, and the caravan encountered no bandits at all. Snow broke out in a strange rash about which Rock only remarked, "I hope it is not contagious." In one small village where they spent the night in a ramshackle temple, Rock read his own scribble on the wall: "Joseph F. Rock. Will I ever come here again? No, never. Jan. 28, 1930."

Snow crossed out the 'no, never' and wrote "yes, again, Feb. 7, 1931." A week later they arrived in Tali, and Rock divested himself of Snow, "a most ungrateful and impertinent individual . . . I paid all the expenses because he would not have had money enough to take him half-way to Tali. I paid for all the escorts, and his parting words were 'I may see you before you go,' without even so much as a word of thanks . . . He is a greenhorn of the first order and an impertinent sponger."

Alone at last, Rock headed toward Likiang, full of misgivings. This time there was no joy in him as he got his first view of the snowy peaks. His old house in Nguluko was drafty and infested with rats; he had rheumatism in his right shoulder. Watching the happy reunion of his Nakhi men with their assorted relatives only heightened his own loneliness and he wished himself in Hawaii or a good restaurant in Vienna. "I shall not complain again in civilization," he promised in his diary. George Forrest was reported to be on his way to Likiang, and Rock was determined not to see him even if it meant hiding out in Muli for a while.

At the beginning of March Rock had a letter from Oakes Ames, an orchidologist and the new head of the Arnold Arboretum, asking him to collect orchids on a part-time basis in southwestern Yunnan and in the Mekong valley; Ames offered $1,000 a year for the work. Rock took offense and replied that he would not consider less than $2,000 as a fee, citing bandits and politics as dangers. He complained that the Arboretum had never really appreciated his efforts in 1926–27 and ran down his long list of "terrible experiences" during that expedition."[8] Frankly, however, now that he was in Likiang, his interest in Nakhi culture began to stir again and he preferred studying native traditions to collecting plants. (A few months later, when he started to worry about money, Rock changed his mind and did some photography and collecting for Ames.) He hired a *dtomba* and resumed his work on the religious ceremonies. The clouds which had covered his spirit for half a year gradually began to part.

At the end of March, dodging an encounter with Forrest, Rock hurriedly departed for ten days of wandering in Nakhi country to the northwest of Likiang. From the moment he hit open camp in the mountains above the roaring Yangtze, he was a changed man, at peace with himself again. He unleashed superlatives in his descriptions of the scenery and lost the oppressive sensation of loneliness. He aimed for Bardar (Bder-dder in his published account), a Nakhi territory in the Y between the Yangtze and the Chungchiang Ho (long. 100°, lat. 27.5°). It was a short trip, bandit free, and every moment of it enchanted him. His objective was to contrast the Nakhis of Bardar with their counterparts in Likiang. The former, not surprisingly, proved to have been "uncontaminated" by Chinese culture owing to their geographic isolation. As a bonus Rock made one of his most delightful and least-known discoveries: a series of limestone terraces and sinter basins in remarkable formations, like an outsided wedding cake, which covered an entire hillside. As he described it later in his contribution to the Harvard-Yenching Monograph Series, *The Ancient Nah-ki Kingdom of Southwest China:* "The basins and terraces have all the appearance of being artificial. Some resemble a terraced, flooded rice-field awaiting planting, save the water is a bluish-white and the banks of the terrace a creamy-yellow, with millions of corrugations, every wavelet a thin shell of lime, of which there must be thousands of layers. The whole structure sounds hollow as one walks over it, but it does not break."[9] The local Nakhis had endowed the marvelous hill with sacred significance, and it figured in their rituals and literature. Rock gleefully informed Graves that he had "discovered a second Yellowstone Park" and would send his account to the *London Illustrated News,* small revenge for his imagined mistreatment. He ended the letter cheerfully "with kindest regards and best wishes and hoping that you are not holding a grudge or enmity against me."[10] His newfound good humor lasted all the way back to Likiang and stayed with him. He thought about Americans rushing around in cities, doing things they did not like doing, while he was precisely where he wanted to be; he felt smug.

ROCK SIZES UP THE MISSIONARIES

Replenished with energy, Rock expected to plunge ahead with his translations of Nakhi manuscripts, but the new *dtomba* balked at explaining the religious mysteries and, after a week of useless pumping, Rock finally gave up and fired him. The weather was gorgeous and, rather than search for a new sorcerer right away, he decided to indulge himself in a short vacation. Not far from Likiang, at Peshui, was a lovely mountain meadow where he had once camped and of which he had dreamed during his bad days in the United States.

Though it had been less than three months since Rock had gotten rid of Snow, he had already forgotten his rules against traveling with white men. Full of good intentions, in April, 1931, he invited Mr. and Mrs. Andrews, Pentecostal missionaries from Likiang, to share his heavenly meadow on a camping trip. The Andrews turned out to be about the worst company he could have chosen for anything. Mr. Andrews, a semi-literate former coal-miner who had been "called" to God's service, perhaps misunderstood Rock's invitation as a plea for spiritual assistance, because no sooner had they pitched their tents than the missionary went to work on the infidel, reminding him in graphic detail what he could expect on Judgment Day if he did not act quickly to redeem himself. Rock put up with this banter for about half an hour before telling Andrews he would not stand for any nonsense and to shut up. Mrs. Andrews was insufferable, "really silly and stupid, and on nearly everything she possesses is written 'Jesus loves me,' even on her eyeglass case . . . I came to think that such is rather a preposterous belief and that Jesus has no choice in the creatures he loves and, if he does, he must have a prediliction for a rather silly and stupid type . . . if the Pentecostal missionaries are a sample." Rock should have known better than to have chosen evangelists as company. After nine years of experience, his opinion of missionaries in general was hardly better than his opinion of the Chinese.

Rock tried to be fair in his evaluation of missionaries and carefully excluded doctors, teachers, and individuals such as Cornelia Morgan from his criticism. But the remote regions in which he traveled had more than their fair share of what he called Holy Roller types—faith healers, evangelists, free-lancers, people who spoke in tongues—but seldom in Chinese or the dialect of their district—and even a man named Brown who declared himself a new denomination and ordered letterheads in Chinese for Brown's Salvationists. He listed the officers of the sect as president, God; vice president, Jesus Christ; treasurer, Joseph Brown.

Rock sometimes wondered if all the crackpots in Christianity had not convened along the Tibetan borderland. The preponderance of bizarre sects, or bizarre representatives of conventional sects, among the tribes of the Chinese West had an explanation. The missionaries, rather like botanists and explorers, had a strong sense of territory and, except in cities where there was enough work for everyone, went out of their way to avoid competition. When Rock arrived in Choni, for example, there was only one mission; sometime later another missionary arrived and asked the Choni prince if he could buy a piece of land on the Tao River and build his mission there. The prince, excessively generous where foreigners were concerned, simply gave the man the land as a gift and said he could cut all the trees he wished for his building. When the first missionaries heard what had happened they complained to Yang that the new man was intruding on their territory; the prince politely reminded them it was *his* territory. The larger, richer denominations monopolized the accessible and heavily populated areas; the churches sent the best missionaries and doctors to the largest missions, usually in cities. Assignments to the hinterlands were considered hardship posts and were distributed to inexperienced or inferior workers. Also, the western mountains were not well known, the dangers and expenses considerable, the tribes difficult to convert, and the returns in Christian souls disappointing. The competition, therefore, was less and the field open to eccentric and independent men.

For every missionary Rock admired there were twenty disasters like Andrews. Rock encountered a few hard-working, intelligent Roman Catholic priests, usually French; a handful of country doctors—one excludes good Dr. Watson, who qualified as a city doctor; earnest people who ran small orphanages and schools in their missions, and William Simpson, the only Pentecostal for whom Rock ever had a kind word. For however much of a nuisance he had been as a traveling companison, Simpson was a generous, devoted Christian who got along extraordinarily well with the Kansu Tibetans. But Rock had met too many hard core soul-savers and had had some terrible experiences with them. He never forgot the time a young couple, faith healers, had called him over to have a look at their baby boy. The infant had amoebic dysentery, and Rock prescribed some emetine, but the parents would not permit him to administer it. The Lord, they said, would decide if the child should live. "You're deliberate killers," Rock shouted in a rage as he left. Ten days later they sent for him again and said: "We've been thinking, maybe the Lord wouldn't mind if you helped with his work." But despite his best attentions, the child died within an hour. "It's the Lord's will," the couple kept repeating, and did not even cry.[11]

Other incidents, though less dramatic, disturbed him almost as much. There was the missionary in Choni who spent his early mornings trading horses and rifles and, after 10 a.m., was too exhausted for anything else; or the missionaries of three denominations on the same street in the same small town who had never spoken to one another; or the one mission family that refused to lend a baby carriage to the second but said they would rather throw the carriage into the river. Such behavior embarrassed Rock, because he knew that many Chinese and tribesmen in the western mountains received their impressions of white men through their observations of a few missionaries. When the missionaries acted like fools, the natives drew appropriate conclusions.

Rock blamed much of the anti-foreign sentiment on missionary activities, and he was not alone in his views. Other travelers in his day shared his criticisms. Eric Teichmann, a British consular official, though considerably less vicious than Rock, arrived at similar conclusions about missionary work in China. The only satisfaction Rock received was his knowledge that, particularly among the tribal peoples, there was mighty little enthusiasm for Christianity. As for the Chinese in general, a foreign devil was a foreign devil, whether he was a representative of God or Standard Oil or, like Rock, an agnostic.

Meanwhile, the picture of China that the missionaries painted for their sponsors in Europe and America was equally misleading and infuriating to him. The missionary's point of view, however humane, was limited by cultural and religious preconceptions which audiences in the West did not necessarily take into account. Professor Fairbank remarked that "in the United

States the picture of Chinese drug addiction, prostitution, footbinding, concubinage, and unspeakable vices excited the morbid, while heathen idolatry and sin appalled the devout."[12] Generous Christians responded with cash and undue optimism. Fearful of being evicted from China and ashamed of the injustices heaped upon the Chinese by the unequal treaties, many missionaries espoused Chiang Kai-shek and his party with professional zeal. "They persuaded themselves," explains Tuchman, "that the Kuomintang, with its source in the Christian Sun Yat-sen, was the sincerely progressive force that would at last end civil strife and bring good government to China."[13] The moral and political views reached vast audiences in the West and influenced public, and eventually official, opinion. A few missionaries, of whom Rock's erstwhile traveling companion Simpson was one, sympathized with communist programs, or at least with programs not blatantly anti-religious. But whatever their leaning, missionary political analyses were dangerously over-simplified.

During the 1930s missionaries in the Chinese interior, regardless of their quality, had become an increasing burden on their governments and the officials responsible for their safety and the integrity of mission property. Communists and bandits —the distinction was not always clear—captured them for ransom with greater frequency than ever. During October, 1930, in the consular district of Hankow alone, twenty-one foreigners, most of them missionaries, were kidnapped.[14] Kuomintang authorities, absorbed in the more pressing problems of civil war, did little to effect their releases. The foreign consuls had their hands full, and missionaries often made their jobs more complicated by refusing to abandon their posts. The State Department of the United States observed glumly that "in response to advice from the Department's officers in China that places of safety be sought, the reply is sometimes made that the persons warned are grateful for the solicitude of their Government, but that danger exists practically everywhere in China, that it attaches especially to the missionary vocation, and that the writers feel a moral obligation to remain at their posts."[15] Washington heard so many tales of woe from China that the State Department had to explain the problems of protecting lives to the parent organizations in the United States in the hope of securing their cooperation.

The results were not encouraging, and too many stubborn missionaries ended as dead missionaries.

Rock was also quite capable of inflicting consular headaches by ignoring sensible advice. Had he always thought of safety first, he probably would have accomplished about half the work he did in China. Though he would have resented the comparison, his logic for traveling dangerous roads to out-of-the-way places was the same as that of the missionaries. He, of course, would never accept the parallel, his contempt for the West China type of missionary being absolute. "Few were as comfortable at home as in the mission field," he noted, "where they had servants, amahs for their numerous children, cooks, and gate keepers, while at home they would have been running around with a pick and shovel and a tin lunch pail. It [was], however, easier to go to West China and convert the heathen." In cases where he did not question sincerity—individuals like Andrews were disarmingly genuine—he remarked wearily that "to talk all day long of Jesus and nothing else becomes boresome, no matter how good a Christian one might be."

Rock determined not to let the Andrews spoil his pleasure in the Alpine meadow and, by imposing rigorous discipline on the conversation and wandering off on his own to collect plants, he succeeded. When he returned to his house, it seemed stuffy and uncomfortable, but he was ready to get down to work. He sent three Nakhis southwest to collect birds for Harvard, found a new *dtomba*, and settled into a peaceful, intense study period which lasted for almost a year. He spent the summer and fall in Likiang and moved to the island of Nyorophu in Yungning for the winter of 1931–32, enjoying the company of his friend the Tsungkuan. The bandits in western Yunnan seemed to have decreased, the people of Likiang had temporarily given up jeering at foreigners; he could not explain either change, but he was freer of anxieties than ever before in China. In July he came down with (self-diagnosed) amoebic dysentery, hemorrhaged badly, but survived in spite of all his fears after a month of bed rest.

He then received a tempting offer from a most unexpected source. "A special delegate from Nanking," he wrote Graves in September, "a Tibetan (well educated) has been appointed to head the new province of Sikang in which Minya Konka and Muli are situated . . . He has arrived in Likiang and has brought letters of introduction to me from several Chinese officials. He is to pacify the province, etc. He has invited me to become his advisor and travel with him through Tibetan country, Tibetan provinces, and the various out of the way places where it would be impossible for foreigners to go otherwise." Flattered by evidence of official confidence and restless after his convalescence, Rock jumped at the opportunity and proposed meeting the Tibetan in Chungtien in October. However, nothing came of the affair. After the initial flush of enthusiasm had faded, he had second thoughts about compromising his political neutrality and getting involved in potentially dangerous tangles. He remained in Yunnan, a decision that eventually proved wise. The Tibetan betrayed the Nationalists when he arrived in Batang; had Rock been with the party, he would have been in trouble.

Rock's unruly digestive system often made him disagreeable but, on the whole, these were pleasant months of accomplishment. His Nakhi studies progressed nicely, and he wrote a rambling narrative about his adventures among the Tibetans, which he sent to the National Geographic in hopes of earning some extra money. But the story duplicated his earlier articles to such an extent that the *Magazine* editors refused it. Nevertheless, he worked prodigiously and with pleasure, untroubled by the outside world.

"Where I live," he boasted to Graves, "we know nothing of depressions; all the people here live by agriculture alone, the crops being good, and everyone raises enough to keep body and soul together, so there is no want. There are no beggars (I have never seen a single Nakhi beggar) and as people buy only what they need and that is not very much, there is no turnover of cash, practically none. For agricultural products and homespun hemp cloth the prices are exactly the same they were when silver was at its highest level. There are no factories, no autos, nobody works for a living—that is, in an industrial capacity—hence there are no hard times. [Of] everything else that hits Shanghai or the east or north of China, may it be war, depression, riots, etc., there is not the slightest repercussion here. We may as well live on the moon and grow our vegetables and raise our meat there, so uninfluenced are we here. Had I not told the people here of the floods in central China they would have known nothing about it. Nobody reads a newspaper for two reasons: first there is none and second the people can't read, and if you want a third, I think they would not be interested if they could read . . . The greatest crisis and the greatest calamity would leave this part of the world in the most blissful ignorance. I have had letters asking me how it is possible for me to live in China during these trying times, floods, civil war, etc., and I really must laugh."

The terrible floods to which Rock referred had claimed around two million lives in central China, a disaster seemingly impossible to ignore in one's own country, yet Americans undoubtedly knew more about them than Rock's hosts in Likiang. Rock himself passed several weeks uninformed of the Mukden Incident and Japanese incursions into Manchuria and, when news reached him, he took it lightly, never dreaming that he would live to see Japanese war planes bombing Yunnanfu.

Rock ingested bad omens with a kind of political innocence generated by his unruffled surroundings and buried himself in work. In December he went north to Yungning to make further investigations of the Nakhi of that district and to visit the Tsungkuan. He studied the ruling family chronicles and the tribal elements of the region and, aside from petty aggravations, was quite happy until he became very sick again. He got through the worst of his illness in Yungning but could not regain his strength and decided he needed professional attention. At the end of February he returned to Likiang just to be closer to Yunnanfu in the event of a relapse.

Either Likiang had changed during the last few months or Rock's mood had turned sour again. As though to welcome him back from Yungning, one of the missionary's dogs bit him in the leg on April Fools Day. The Likiang Nakhis, whom he had in November likened to "children of nature [who] possess an innate dignity and courtesy of which many a European could take a lesson," had by early April became "hostile and mean." "Communism is rife here," he wrote, linking communism to anti-foreignism. "The students are the worst offenders and they are instilled by their ignorant teachers." When his Nakhi servants went into the streets they were called "running dogs of the foreigners"; they remained loyal to him because he treated and paid them well, but they suffered the contempt of their peers. Rock took even more abuse.

The difficulty of maintaining one's composure, much less one's dignity, in the face of such verbal assaults was illustrated in the writings of Major H. R. Davies, an Englishman of Victorian mettle, who traveled extensively in Yunnan at the turn of the century in search of a link between the Yangtze and India. The persistent cries of "foreign devil" eventually became too much for him: " . . . Watts-Jones and I had walked on a little ahead," he recorded, "and passed five or six Chinamen sitting by the roadside. Directly we had passed, we heard shouts of *yang kuei-tzu*. We turned back at once and our friends instantly dispersed in different directions. However one of them was a slow mover, and we soon caught him and gave him a good thrashing. This, however, turned out not to be the man who had actually shouted, though I have no doubt his licking did him a lot of good. So we set to work to search for the real culprit, and soon found him hiding in a wet nullah. After a short interview, we left him begging for mercy, covered with mud and water, with a bleeding nose and various bruises all over him, so perhaps he will be more civil in the future.

"This expression, *yang kuei-tzu*, has come to be the usual name for a European among the Chinese, and I have heard it used by ignorant countrymen in out-of-the-way places, evidently without realising that there was anything insulting in it; but when shouted out in this way as one passes, it can only be meant as an insult, which it is one's duty to resent."[16]

Thirty years later, Rock was as much, if not more, insulted, but, though often tempted, never once replied to an insult with a blow. Lacking the supreme self-confidence of imperial superiority with which Davies and his ilk were blessed, Rock endured his ordeal in frozen silence. Only once, after an accidental conversation with amiable high school students, did he try to achieve some insight into the situation. "I came to realize that perhaps my own attitude—seeing in every gesture an insult—had much to do with their attitude toward me," he admitted, but he was too set in his behavior and thinking to alter that attitude. His bad health did not improve his temper. At the beginning of April he received a letter from Thomas Goodspeed, head of the botanic garden at the University of California at Berkeley, authorizing him to make a botanical expedition in the Mekong and Salween valley and into Muli. He was in no condition for long trips but could not afford to turn down the offer so, instead, sent a dozen trained Nakhi collectors in scattered directions to do the work on their own. He, meanwhile, repaired to the French hospital in Yunnanfu for treatment.

The familiar caravan route to the provincial capital, which normally crawled with bandits, was momentarily peaceful, but the long arm of Lung Yun was visible everywhere. It seemed to Rock that more fields had been planted in opium; poppies occupied a good 30 percent of the arable land along the trail. Owing to his physical weakness, Rock made most of the journey by chair, a style of travel which attracted attention to his foreigness and brought more taunts than usual. Entering one town not far from Likiang, "the first person we met was a baby of about four years of age, and the first thing that youngster did was to hurl an insult at me. One is perfectly helpless, but it makes one mad and humiliates one before the servants. I could have killed that kid!" Once, to ease the sensation of helplessness, he reported some rude soldiers to a district magistrate knowing all the time that nothing would come of his complaints. As insults heaped upon him and he passed fresh scenes of misery and filth, his old hatred of the Chinese and what he thought of as their indifference revived and he repeated the indictments of the year before when he had traveled in the opposite direction with Snow. By the time he reached Yunnanfu, all the beneficial effects of the western mountains on his spirit were undone.

He was, after all, delighted to be among Westerners again; one could more easily insulate oneself against anti-foreign assaults in the company of fellow victims. Old friends welcomed him. The Pages (he, a Standard Oil representative) had a motor car. Rock rented a house with a screened verandah near the canal and felt so much better that he wished he had gone botanizing with his men in the Mekong and Salween. As before, the few days of well-being passed quickly, and Rock soon entered the hospital with a new seizure of abdominal pains.

MOUNTING PROBLEMS OF CHIANG KAI-SHEK

By 1932 China had had five years of Chiang Kai-shek and the Kuomintang, but the record of domestic activity showed clearly that the Nanking government was a long way from controling the vast and ailing country. The best that could be said was that Chiang had come closer to achieving national unification than any political aspirant since the demise of the Manchus in 1911. This accomplishment, however, was small consolation to a regime beseiged with enemies from both within and without.

Much of the responsibility for the triumphs and failures of the Nationalist Government during the pre-war years must be assigned to a single individual. Chiang Kai-shek has been served inordinate portions of praise and blame for his role in the development of China, and has been discussed with as much passion as any political figure in this century. To his critics he was a monument to militarism and egocentricity; to his partisans who, one notes, have dwindled of late, he was a hero of mythical proportions. In 1932, however, it was still possible to occupy middle ground on Chiang. Rock, for example, wrote him off simply as another warlord which, in some respects, showed good judgment because Chiang was a product of the warlord tradition and his rise to power had been effected in warlord fashion, by a series of military victories and alliances. But he was certainly no ordinary warlord. He had great courage, a fine military instinct, a deep love for his country, and, within his limitations, good intentions. He had been a young revolutionary, and once in power he attempted needed modernizations such as currency reform and improved communications. But like many soldier-statesmen, Chiang had deeply conservative values. His cultural and moral concepts were anchored in the Middle Kingdom. Hindsight makes it evident that he misjudged the peasant masses which constituted the backbone of the nation he so desperately tried to rule. He failed to grasp the obvious fact that China's greatest resource, in peace as well as in war, was its miserable millions of people. Maybe the very simplicity of the notion was precisely what caused him and most of his contemporaries to overlook it. Mao Tse-tung, after all, spent years convincing his communist colleagues that a peasant revolution could succeed.

It is easy now to list Chiang's shortcomings and to disregard the staggering complexity of making political order in China in the 1920s and '30s. He realized immediately that one of the country's biggest problems was indiscriminate militarism. As Rock had so often witnessed in the west, the hordes of soldiers who roamed the country were an enormous drain on its energies. Chiang tried to remedy the situation in the same manner as Yang Sen of Szechwan, by asking other generals to disband their armies and promising, in the indefinite future, to reduce his own. Unfortunately, he did not offer these generals any compensation for the loss of their forces; he was unwilling to share power in government. In 1929, therefore, he lost the loyalty of Feng Yu-hsiang and the Kwangsi generals. He had to fight both and, meanwhile, the military burden upon China increased rather than diminished.[17]

Apart from dissatisfied generals and other insurgents with personal ambitions, Chiang had to contend with communist rebels whom he feared more than any warlord, possibly because they did not conform to his conventional idea of enemies. The Kuomintang expulsion and purge of communists from its ranks in 1927 had scattered them in various directions, but the movement had unanticipated regenerative powers. As Chiang began to learn, after a brief period of seeming defeat, the communists returned with renewed vigor, as though having gained nourishment from the attacks upon them. During the summer of 1930, under the ideological guidance of Li Li-san, the newly formed Red Army attacked Wuhan, Changsha, and other industrial centers in the middle Yangtze valley. Though the action failed, it alerted the Nationalists to the necessity of eliminating the communists. In December, Chiang ordered the first bandit-suppression campaign. (He would not dignify his adversaries with a political name.) When it failed, a second and equally unsuccessful campaign was launched between February and June 1931; Chiang personally commanded a third effort and, by the beginning of September, 1931, thought he had the enemy licked. In the middle of the same month, however, the Japanese attacked Manchuria and unwittingly gave the communists a new lease on life.

In the opinion of many historians, Chiang then committed his fatal error. Instead of using Japanese aggression as a rallying point for national unification, he chose not to resist Japan before he had ridden himself of domestic dissenters. The consequences of this decision, nationally and internationally, were grave. Japan, which had been divided at home on the Mukden effort, was encouraged by success, and expansionist elements in the military structure grew bold. Most of the bedraggled Chinese communists retreated south to Kiangsi to regroup after Chiang's third campaign; they adopted a provisional constitution and set up a provisional government. In February, 1932, that government declared war on Japan. It was not until 1937, when pushed to the utmost limits of endurance, that the Kuomingtang engaged Japanese troops. The Nationalist government did not even declare formal war on Japan until the day after Pearl Harbor! Instead, during those years, Chiang pursued communists and other opponents with stubborn energy. The result of his strategy was to push those who resented the Japanese presence in Manchuria into the arms of the so-called bandits and to set in motion the epic Long March.

Momentous decisions by Chiang Kai-shek, Japanese in Manchuria, and ideological squabbles among communist leaders had little impact on the western provinces. It was easy for foreigners in Yunnanfu to be unimpressed by rumors and newspapers though Rock, after his 1926–27 experiences of the Feng Yu-hsiang army, was inclined to be excessively suspicious of communists. Until the Long March and, in many respects, until Chiang moved his government inland to Chunking, the western provinces maintained their innocence. Lung Yun's nominal alliance with the Kuomintang and his selfish military adventures did not alter Yunnan's practical political sovereignty. The province was still full of opium, bandits, and truculent tribes.

Clubb's analysis of conditions in Szechwan provides an illustration of the limits of Nanking's power. "That province, he wrote, "secure behind the forbidden mountain ranges that enclosed its fertile Red Basin, was the prime example of nonconcern with Nanking's rule. Three old-time militarists bickered endlessly among themselves for spheres of influence in northern Szechwan [one of them was Rock's erstwhile protector, Yang Sen]. In the rest of the province the struggle was between Liu Wen-hui and his uncle, Liu Hsiang. The province groaned under the burden of an estimated 200,000 troops; none of them under Nanking's control, all hungry for territory on which to feed. Not only was Szechwan beyond Nanking's authority and its laws, but there was not even *local* control." The author was not even counting mountain enclaves such as Muli or wild Lolo country.

In Kansu the Kuomintang had somewhat more authority by 1932, though its presence did not exorcise independent warlords or the intransigent Tebbus. But the Nationalists were no more of a blessing to the people of Kansu than the Kuominchun had been. Clubb offered a list of forty-four taxes, of universal applicability, to which the people of the province were subjected. There were taxes on food and clothing; wheat bran, cereals, a flour shop; skin overcoats, hemp shoes, stockings, military clothing, and alteration of uniforms. Taxes on kettles, kindling, houses, "wealthy house," bedding; also on expenses incurred in changing and repairing defences, special loans, merchants' loans, communications, soldier enlistment, road building, trestle work; hogs, army mules, horse-fodder, even "purification of countryside." On top of all this soldiers, Nationalist or private, considered it proper to help themselves to food, livestock, and grain from the peasants. Kansu had also suffered extensive drought and famine in 1929 and the ravages of the Moslem rebellion.

Knowing the western provinces as he did, Rock found it difficult to accept Chiang's pious proclamations of national unity. He never believed for a moment that the Nationalists either could or would solve China's problems. And while he loathed and distrusted communism he could see, as reflected in his statement to Snow, its potential for attracting millions of peasants, though not if it were handled Feng Yu-hsiang style. Rock hoped, contrary to all evidence, that an enlightened political force would materialize and reach the peasants before the communists. "In many ways I feel sorry for China," he wrote, "but mainly when I see how foreigners want to push in and open up the country and force industrialism on an agricultural people. Her sorrows are only due to dishonest officials and utter mismanagement, to dishonesty and . . . ignorance. Give China a real decent government, a *plain* education, teach it hygenic living and prevent the enormous death rate by introducing cleanliness and Western medicine, and reduce the birth rate by making them give up their ancestor worship, which causes them to breed until one or two sons are born. In the meantime come four or five girls who are considered nuisances but must be married at an early age. Otherwise the people should be left alone. China is bound to outlive any so-called highly civilized and highly industrialized state as she has outlived Egypt, Greece, and Rome . . . The Chinese will never want the things Europeans pride themselves on possessing, and it will be an evil day for China when its people come to that state when they must have the useless things we call necessities."[18] Surely Rock meant well, but China was not a problem to be solved simply or gently.

Rock was discharged from the French hospital in Yunnanfu, apparently much improved. He went back to his comfortable house and his work on the Nakhi manuscripts; at the end of June, 1932, he tentatively began a dictionary of the ancient Nakhi pictographic script. The summer passed calmly because, when his work went smoothly, Rock was not so easily riled by street hecklers. Even the presence of the official execution grounds behind his house did not bother him very much. He satisfied the morbid side of his curiosity by witnessing a few executions and then conditioned himself to ignore the unmistakable volleys of gunfire. "We are one happy family here, and there is nothing that mars our peaceful existence," he recorded on July 27, 1932. He entertained transient Westerners grandly and spent pleasant hours with Watson, the Pages, and other foreign residents. He began to accept his change of life-style and, because it was impractical and expensive to keep him, parted with the horse he had used on many expeditions. "I came to realize that the end of traveling in the interior has really come," he admitted. He felt sad but did not give way to depression and suicidal seizures. Also sad was the news in one of the foreign newspapers that Simpson had been murdered by Moslem bandits in Kansu. In spite of their differences, Rock had respected his approach to the Tibetans "He was a real genuine type of missionary who brought no foreign ways and customs to a strange land", he wrote Graves; also, he thought it bad form to speak ill of the dead. Otherwise, as he said, nothing much happened during the summer.

Ever since his hospitalization, however, he had been troubled by headaches, and around August they became quite fierce and frequent, so he applied to the doctors for a diagnosis. They told him he had suffered from a "transient stroke" and that he should go down to lower altitudes; the advice, ironically, came a few days after he had been wondering if he might not feel better at higher elevations and had mused about moving to the Indian Himalayas. He now felt settled in Yunnanfu and could not think of any lowland city in China, because, for the moment, he definitely wanted to stay in China, which appealed to him. But the legitimate excuse for travel excited him and he decided to make a brief trip to China's east coast and look over the possibilities. He ordered two Nakhis to prepare for the journey.

The French Railway and boat brought them to Hong Kong by September 25. Rock felt terrible and went to the hospital for five days, carefully double-checking the Yunnanfu diagnosis like the dedicated hypchondriac he was; but none of the doctors was very helpful. Hong Kong was decidedly too Western for him and, as soon as his health improved, he left for Shanghai. He had never liked Shanghai, Sin City of China, and did not like it any better now. He heard Snow was in town; looked him up, and settled old misunderstandings. It was then that Snow, with the very best of intentions, misjudged him and took him to the Rose Room and profoundly offended his puritanical soul. Poor Snow could not do anything right by Rock. Otherwise, in every way, Shanghai was a jumble of the worst elements of East and West, a visual and intellectual study of contradictions. For Rock it had only one redeeming attraction: the bookstores. He spent the better part of three weeks browsing and buying books on western China.

Peking, then Peiping by decree of the Nanking regime, was a different story, a city of charm and dignity which, despite a sizeable foreign colony, had remained relatively untainted by Western ingenuity. The bars, brothels, and coarse mercantilism that characterized Shanghai and other large treaty ports were scarcely noticeable in the ancient, though temporarily demoted, capital. Peking was aesthetically beautiful and intellectually alive, something that could not be said of any other Chinese city. There were libraries, museums, and universities, both Western and Chinese; traditional Chinese scholars, gentlemen of unworldly refinement and impeccable manners, perused old texts; young students, many of whom had been exposed to Western education, engaged in hot political debates; foreigners lived so graciously, and cheaply, that many transients settled down in the fashionable Western Hills. Rock, on his second visit, was absolutely enchanted. Through contacts with various official and university people, he traveled in rarified circles where Chinese dignitaries treated him with mannered courtesy; even Chinese in the streets seemed less hostile than in Yunnanfu. Rock saw his first "talkie," a Gothic monster business, spoke with scores of interesting people, and imagined he could live quite happily in Peking. After a week, however, he started his return voyage to provincial Yunnanfu which, after all, was home these days. Back to Shanghai and Snow, whom he had forgiven; Hong Kong, an amusing side trip to Macao, then Hanoi and the train, and on to familiar Yunnanfu. As it drew into the station, the train ran over and killed a coolie.

Rock found everything in order and resumed his old routines. His health was still mediocre, but the headaches came less frequently. The Nakhi men who had been out collecting for Berkeley arrived between Christmas and the New Year with twenty-six boxes of plants and seeds, and Rock congratulated them for their work. He went on with his translations of Nakhi religious ceremonies, putting the dictionary aside, spurred by a sense of progress and the possibility of a publisher. The next half year passed quietly. He received a letter from Washington with the news that Ralph Graves had died and, no matter how badly they had quarreled, Rock hated losing another piece of his past. The young British Vice-Consul, a witty man who made social life in Yunnanfu tolerable, got peritonitis and died in the hospital, much to the regret of the English-speaking community. In April, 1933, Rock fired his *dtomba* for improvising the books he did not know and thereby creating scholarly havoc. Rock sent back to Likiang for a replacement and sorted out the mess while awaiting his arrival, accepting the situation calmly. He was in a philosophical mood and now read Spinoza and Nietzsche instead of Dickens before he went to sleep. Brooding, he observed, "leads nowhere."

Six months is a long time for a man who feeds on excitement and change to stay in one place buried in his books. By summertime, boredom and malaise insinuated themselves into Rock's life, and his stomach began to misbehave again; like the chicken and the egg, it was hard to say which was cause and which was effect. "My life has been a lonely one," he complained in his diary in June, as he started his downhill slide into despair. A week later he was horribly ill, and one of the doctors had to sit up with him all one night. He would not watch his diet. Now that he was out of buttered tea country, he had no excuse for not eating properly, but he could not control his longing for rich foods. He spent an uneasy ten days in bed at the end of June. It was the rainy season, and the weather was wet and oppressive.

Then the floods came to Yunnanfu, only three days after Rock had recuperated from his attack. Kunyang Lake swelled and poured its contents into the canal which overflowed its banks, washing away thirty houses on the first day. By Chinese standards it was a tiny flood, scarcely worthy of the name disaster. The fields were under water and the prices for vegetables doubled. Homeless people stared vacantly at the spot where their houses had stood. They had been through floods before and would wait until the water receded to rebuild on the same endangered site. Darwin had made a similar observation almost a century before as he surveyed the rubble of an earthquake in Chile.

Rock, however, was disinclined to humanistic speculation and watched the whole process with more irritation than sympathy. The Chinese were on his nerves again. In his low periods he was always more sensitive to their jibes, and there had been a few really unpleasant incidents. "I have been taking my daily rickshaw rides on the circular road, but every Chinese one meets sneers and grins at one, and it is no pleasure to go out in this place, for all one gets is insults. The other day as I passed along in a rickshaw, a Chinese spat at me . . . " The city crawled with thieves, and the almost daily executions did not deter the desperate. Disorder filled the air, and there was a rumor in the city that there would be a revolution. Likiang, Rock thought, would be different; he forgot his miseries there the year before. He also considered establishing himself in Peking.

Misfortune piled upon injury and insult. Rock had tried to improve his financial position by speculating in foreign currencies. It was a bad time to play the international money market without being well versed in finance, and he had been unlucky. Not believing in fate or his own bad judgment, he cursed his Washington bank. In mid-August when he received the announcement of the death of the Tsungkuan of Yungning, he wept. Deprived of his dearest friend in China, his longing for Western civilization quickened. He continued his translations, plodding through book after book with, finally, a good *dtomba*. He had piles of manuscript, and an English publisher had expressed some interest. Rock had no plans for the future, no destination, no project. "My life moves in cycles of thirteen years," he decided. "I was born on a 13, I lived thirteen years in Hawaii, and have been absent just thirteen years. My last visit to Vienna . . . was also thirteen years ago . . . " He resolved, as much on the basis of superstition as of reason, to leave Yunnan and go to Europe.

This time Rock determined to get out of Yunnanfu once and for all. He packed eighteen trunks full of books and belongings and shipped them to Haiphong. He closed his affairs with the consulate; "and I still have to go once more . . . re that wretched income tax for the government to squander on useless extravagances [such] as changing consuls three times in three months in an unimportant place like this. . . . I went to see them the other day and they all sat around and loafed and said, 'Glad you came to give us something to do.' " He gave notice to his Chinese landlord. By early October he had finished his translations and, having disposed of his worldly goods, had only to fix the day of departure. The thought of permanent separation from his Nakhis pained him, but he tried to reason it away: "They are kind, quiet, simple-minded, of a childish innocence, therefore, almost lovable. I like them but I want intellectual companionship. . . . With these men it is like always being surrounded by children." The rationalization did not succeed and about four days before leaving he approached Li Ssu-chin, his favorite Nakhi, and suggested he travel along to Europe. Li declined and said he wanted to return to his wife and mother in

Likiang. At the last moment, the Nakhis made parting easy. When he distributed his leftover clothing, shoes, suitcases, etc., and gave them their bonuses, the "simple children" made clear that they had expected greater rewards. Rock was hurt and disappointed; he was as innocent as they and sorely wanted to be missed for himself. They left him few illusions, and he regreted only leaving Li.

Rock started for Hanoi on October 19, 1933; by the 21st he was already homesick. He missed his greedy Nakhis and his house near the canal and the execution grounds. He was still unsure where to go, only that he would "like to live where there are no changes." In Hanoi he ran into the Pages—they formerly of Yunnanfu and the motor car—and dined with them along with a woman who made him think of a "pinched bat" and who poisoned his evening. After dinner, alone in his hotel room, he wondered where to go. His nephew, Robert Koc, was very much on his mind. Several times during his loneliness of the last few months he had toyed with the idea of sending for him. When Li Ssu-chin turned down his offer of a European trip, Rock faced a future without familiar servants or company, however inadequate, unless he returned to Yunnan. Young Robert might fit the role perfectly and fill the vacancy in his heart, yet he wavered between desire and dread. Never in his life had he solicited emotional involvement, and he feared success with its implications of restraints as much as he feared disappointment or failure. In Hong Kong he made his choice suddenly, in favor of Europe; he cabled Robert to meet him in Italy. One major decision at a time being quite enough for him, he left all his trunks in storage in Hong Kong, where he could collect them easily in case he returned to China. He then proceeded to Shanghai to meet his ship.

The *Conte Verde* was filled, according to Rock, with "fat, painted Standard Oil women and their oil-selling husbands . . . [and] fat Jewish diamond sellers and their dubious women or perhaps wives," cheerless company for a month-long ocean voyage. Lacquered toenails, then very much in vogue, symbolized for him the depths to which Western civilization had sunk. He kept to himself in disgust and, in the privacy of his diaries, indulged in some unbecoming anti-Semitic remarks.*

The ship put in briefly at Singapore, and Rock rushed over to the Botanic Garden to look at the chaulmoogra trees he had left there in 1921. The other ports of call, colorful, noisy, teeming with life, depressed him. The waterfronts were always crowded with avaricious salesmen pressing their wares upon him—everything from indecent postcards to indigenous art. The closer he got to Europe, the greater were his misgivings. He landed in fascist Italy on December 8. The next day he wrote with conviction: "Europe, no, it has no more appeal for me."

He was in Venice for the first time in thirty years, spending money savagely in revenge for the days when he had nearly starved there. Robert should have met him immediately, but there had been a mixup, and it took them three days to find each other. They met, almost by accident, in the morning on a sidewalk near Rock's hotel. Robert was quiet, thin, weary, and ragged, and Rock's first instinct was to buy him a new outfit and, at least, improve the surface. By evening, Rock was wary: "Now as to Robert," he wrote in his diary, "poor boy, he seems to be weak with a little cough. He has a kind look but certainly will not set the world on fire." Hints of disappointment. Nothing revealed more surely the depth of Rock's longing for someone to love than this gentle criticism; it was the closest the ever came to giving anyone the benefit of his doubt. His forebearance lasted three or four days, but with every hour it became harder for him to sustain. He cast about desperately for excuses for Robert's apparent weaknesses, which seemed to him both of body and of character, but the best he could do was keep telling himself that his nephew was a good boy. He compensated for Robert's limp-

---

* In all fairness I must report that when I mentioned this side of Rock to Henry Corra, an Austrian who knew him for many years, the latter was astonished. The late Mrs. Corra was Jewish and Rock had always been very fond of her. Nevertheless, entries such as " . . . the typical Jew, short of stature, bow-legged, eagle-nosed, curly hair, and clothes such as only a Jew would select, hence make more conspicuous his racial characteristics and features," are not what anyone could call unprejudiced. He had, of course, been raised in a part of the world where anti-Semitism was a popular pasttime.

ness of attitude with excessive activity, rushing from Venice to Genoa, to Nice, to Monte Carlo, and he expressed his disenchantment by explosions of hostility. His unpredictability, added to his inclination to treat Robert like one of the Nakhis, confused and frightened the young man. Both uncle and nephew were under great strain, the one desiring to be pleased, the other desiring to please. Only when Rock admitted to himself that Robert was entirely unsuitable for any of his needs did the tension ease. He announced his decision with uncharacteristic delicacy, explaining that he really needed a secretary and companion who had mastered the English language. Since he had done without one for years and knew beforehand that his nephew had not studied English, it was a flimsy lie, but it worked. Robert believed it gratefully and promised to dedicate himself to the language in case his uncle needed him sometime in the future. Rock let him have his illusions—a little English would be good for him anyway—and the atmosphere relaxed a little. Though his hopes could not have been more thoroughly dashed, Rock was conscious of more relief than pain. No matter how lonely his life, he was used to it, and tampering with its emotional structure had made him nervous. They traveled more easily now, bound for a Viennese Christmas.

Just about this time the *Honolulu Star-Bulletin* disclosed that "Dr. Rock Ends His Long Life in the Wilds; Famed Botanist Goes to England; Too Old to Continue His Work."

X

THE YEAR OF THE LONG MARCH

Vienna touched off memories like a series of small explosions in Rock's brain, and he chased after his childhood through the city's streets and parks. He discovered that Count Potocki's grand *Winterpalais* on the Schottenring, home of his youth, had been sold and was in the throes of being remodeled according to the specifications of an Italian banking firm which would soon move onto the premises. The Hotel Sacher was not quite what it used to be, or what he imagined it used to be. He stayed at the Grand Hotel. "Something seems to be missing," he noted in his diary; maybe imperial splendor, yet he was filled with affection for Vienna. He sought faces from the past—his Latin teacher from *Gymnasium* days and a former schoolmate who had grown ugly, walked "like a duck," and had no interesting conversation for him. The opera and theaters were excellent, and he enjoyed them with the appetite of a man who has been long deprived of his culture. And then there was family and Christmas Eve with his sister Lina and the three boys and even Rosa Wetzl, the adopted sister Rock had resented so much as a child. Lina baked strudl and decorated the tree; Uncle Rock bestowed lavish presents and remembered aloud until his eyes filled with tears. Always the fatalist, he reckoned this would be his last Christmas Eve among his kindred.

The fact that he had earned a scholarly reputation within a small, distinguished circle of Viennese savants did not at all diminish Rock's pleasure during his visit. He had corresponded with Handel-Mazzetti, the botanist and explorer who had worked in the same general part of China, including Likiang, between 1913 and 1918, and who bore the title of Baron; now they sat together in the dining room at the Hotel Sacher over *Naturschnitzel* and conversed amiably. Rock received invitations to address botanical audiences and obliged with glee. Nothing soothed the ego of the servant's son more than to be accepted in intellectual quarters.

If any Western place could have captured Rock then, it would have been Vienna. But Vienna did not try. There was nothing for him to do there except, possibly, to join forces with Handel-Mazzetti in a phytogeographical study of western China, and cooperation on that level never entered his mind. He stayed two and a half weeks in his birthplace and then tore himself away. Departure was not easy. The matter of Robert still troubled him—Robert made a better impression on home ground—and the disappointment lingered. But he left his family, such as it was, behind on the railroad platform on Jaunary 9, 1934, on his way to London.

He stopped a few days in Paris where he had a bad hotel recommended by a fellow train-passenger. Prostitutes kept accosting him on the streets. He became bad-tempered again, unhappy with Europe, and impulsively cabled $300 to Miss Scharten in Likiang to pay the expenses of his Nakhi men for their journey to Yunnanfu; he would wire the date of his return later. Then he made methodical rounds of Paris attractions: first, on his fiftieth birthday, to Les Invalides to see the tomb of one of his heroes; next to the Louvre which he left, like any other tourist, wishing he had more time. But he had to push on across the Channel.

As soon as he arrived in London he hauled his Nakhi manuscript over to the publishing house of Kegan Paul. The editors had been theoretically interested in Rock's translations of religious ceremonies without realizing they were dealing with a compulsively thorough scholar. When he materialized in their presence with the bulky opus clutched under his arm, they were aghast and explained apologetically that they could not possibly handle it. *Rock appears to have been not at all distressed by this announcement, accepting the publisher's rejection as a compliment to his work rather than as a comment on its obscurity. This had been the one practical excuse for his trip to Europe and it had failed, but he did not care. Instead, he made the best of England. Mrs. Frederick Muir, the widow of his old friend and landlord from Hawaii, was there, and they took long walks together in the English countryside, reminiscing. London was full of botanists and horticultural enthusiasts who knew about his work, so Rock was much in demand and could choose his society. He visited Rothschilds—who could resist a Rothschild?—first Walter, the ornithologist, "a nice old gentleman and scholar." Though disinherited by his father, Lord Walter had a fantastic zoological museum, a library, and a Gainsborough on the wall; he was intrigued by China and Oriental politics and questioned Rock while tea was served by discreet liveried servants. Rock was understandably devastated yet felt very much at ease in the sumptuous surroundings. Lionel Rothschild, the cousin, was another matter, "the real Jew . . . no scientist, but a man to whom money is uppermost . . . " He was also a lover of orchids and rhododendrons and had some of Rock's introductions among his huge collections, including some conifers sent from Likiang. Nevertheless, Rock's ethnic preconceptions were unalterable.

He called on Dr. Watson's brother and, on the whole, found very pleasant company in England, but he grew more and more homesick for Yunnan. English cooking and English weather depressed him. "The English are welcome to their dreary climate which has made them cold and distant," he said. London disagreed with his health. On the last day of January he nearly col-

---

* Portions of this manuscript were eventually published in the *Bulletin Ecole Française de l'Extrême-Orient,* vols. 37 (1) and 39 (1), in 1937 and 1939, respectively; in the *Journal of Oriental Studies,* Catholic University of Peking Monograph Series XIII (1), 1948; and by the Istituto Italiano per il Medio ed Estremo Oriente in 1952. The work as a whole was the foundation for Rock's dictionary of the Nakhi pictographic script.

lapsed and, a few days later, had to be taken to a hospital. Though the accommodations were practically medieval—there were no water closets, and the patients used commodes—the medical equipment was first rate. The doctors subjected Rock to extensive tests and produced the first correct diagnosis of his intestinal condition: he had an obstruction for which they had no absolute cure without surgery.

Ten days of lying in the hospital under the scrutiny of an enormous X-ray machine gave Rock time to think. He had heard much talk in Europe about fascists and war. From his prone position, the Western world seemed to have gotten itself into an absurd mess. "The white man has been and is the curse of this world," he complained. "He it is who is the snake in paradise. He created all the machines of war which make it possible for death to conquer and embrace men by the millions. And at the same time he created, or wishes to create, a thing [the X-ray machine] to counteract the horrors or ameliorate the frightfulness of his terrible death machines. What folly, what folly! But he is more successful in destruction than in preservation. And what does he preserve in the end? Personified misery: legless, armless men, men who can never breathe properly for they have inhaled the poison of [their] own making . . . Where is the adage 'love thy neighbor as thyself?' It is an empty phrase bellowed forth in cold churches once a week by men who, instead of condemning the weapons of murder, drag God or the name of God down into the mire and bless them by invoking his name." Europe had been a sobering experience for him. From there, China and its amateurish provincial wars seemed like a peaceful haven.

He decided against an operation and sailed for New York at the end of February without any regrets. Much to his annoyance, the ship was full of German Jews. Rock moved quickly through the United States, enjoying himself for a change. Sargent's wealthy widowed daughter, Nathalie Potter, was not noticeably suffering the effects of the depression and entertained him grandly in New York. In Washington he met old acquaintances, particularly E. A. Back of the Department of Agriculture's Plant Quarantine Station. He had no cause for quarrels with the National Geographic Society. Now that they did not

have to deal with his quirks, the officials there received him cere-
moniously. He proceeded to Chicago and then on to San Francisco and Berkeley, where Goodspeed expressed delight over the
results of the expedition he had supervised. Wherever he stopped
he was flattered, entertained, and questioned; he was, accordingly, at his charming best. His ship to the Orient stopped in Honolulu, but he did not interrupt his journey and only had time to
call on a few friends who, having taken December's newspaper
article and letters Rock had written to Lyons for granted, expressed surprise that he was returning to China.

PORTENTS OF CHANGE IN CHINA

Rock leaned over the ship's railing, peering through the
mist at Japan's Inland Sea. It was Sunday, May 6, 1934, and
about two and a half months earlier he had been in London. By
now he was immune to cultural shocks and moved easily from
one continent to another. He was equally at home or equally al-
ientated—depending on his state of mind—in Asia, Europe, or
North America. Leaving Japan, his thoughts were far from Nazi
Europe. "Adieu Japan," he wrote. "Somehow or other my sympathy goes out to its frugal, hardy, hard-working people but not
to its militarists; they are the products of Western civilization.
They thought they must adapt the vices of the Western world
war to protect themselves against it. The white man has been
their teacher. He forced them by threats to open their land and
now by threats they are keeping him out—not only out of Japan
but they are extending the threat to include Asia.

"I am afraid the white man has run his road and will have
arrived at [a] suicidal cliff. . . . I would like to live another
fifty years to see what will have happened to the Orient." Rock's
few days among the Japanese confirmed the fears he had expressed during his conversation with Lord Walter Rothschild in
England. "We discussed the 'Yellow Peril' and, of course, came
to the conclusion that a Yellow Peril exists only from the Japanese side." No Chinese "indifference" there!

In the excitement of returning to China, Rock banished
somber thoughts. He arrived in Peking with the vague intention
of renting a house there. Snow was in town, and Rock was ready

to forgive him whatever he thought should be forgiven; Snow, innocent of his offenses, was just his good-humored self. They lunched together and spent an afternoon house-hunting in the Western Hills. Rock also had business on his mind. He carried a manuscript on the history of the Nakhi kingdom to Henri Vetch, a Peking-based French bookseller and publisher whom he had known for a long time. Vetch looked through the pages and recognized in the bulky disorder a work of cultural merit in desperate need of editorial attention. Rock said he could not afford a professional editor but, if he were living in Peking, he would be able to work directly with Vetch preparing the manuscript for the printer. The publisher, charmed by Rock, and, unlike some of the National Geographic Society editors, never having endured the agony of manipulating Rock's prose, took him out to look for houses again. But the selection did not please Rock, and he was anxious to get back to Yunnan. He decided, finally, to spend the summer in Yunnanfu and return to Peking after the summer foreign residents had vacated.

Having solved that problem, he settled down to serious sight-seeing because he had left much undone on his earlier visits. "In the Chinese city," he noted, "there is chaos, no order, everyone rides, walks, and loiters in a fashion unknown in an orderly European city. The fronts of shops or facades of houses, if they can be called houses, are irregular to a degree like orderless junk thrown about or bamboo poles of various lengths stood up haphazardly. Strips of cloth, white or blue or black with Chinese characters painted on of all sizes, hang from frameworks. The sidewalks are littered with the cheap trash of peddlars. Mangy dogs roam about snarling at each other when confronted by a clean-scraped bone. Beggars covered with rags and filth, worse off than the dogs, try their best to obtain a copper or some rice for the next few hours of existence." In all this misery there was a kind of vitality that touched even an old China hand like Rock. The scene as a whole was picturesque; only by observing individuals could it be stripped of its superficial charm, but few Westerners could look upon a Chinese as a genuine individual and Rock, while recognizing great poverty, viewed the tableau romantically, more resentful of the occasional intrusion of foreign elements than of the desperation and death surrounding him.

Old Imperial Peking, the museum, the libraries, the Great Wall, and especially the Ming tombs enchanted him, but more evidence of Western influence than he had noticed on his last visit injured his sense of historical dignity. One incident in particular elevated his blood pressure. He went to see what had once been Peking's most sacred temple, the Tai Miao where emperors had prayed, and found Chinese laborers scurrying about with hammers and nails preparing for a fair which would show off machinery, agricultural tools, canned goods, and other ingenious paraphernalia. The Western tourist was horrified. " . . . I wished I could take a whip and, imitating Christ, I would like to drive them out of this hallowed precinct with all their claptrap of cheap foreign imitations. What must the old juniper trees, hoary with age, think to see coolies sitting beneath them sipping cheap banana soda water."

Yet Rock still liked Peking and considered it a likely home for the future. He stayed for two weeks and then went on toward Yunnan. The train ride to Shanghai thrust him into a militaristic milieu which had not been evident in the ancient capital. Every railway car was guarded by a pair of soldiers with Mauser pistols. Soldiers, railway guards, and police with rifles and fixed bayonets protected the train at every stop. Rock noticed other disquieting signs. "There are some militarists on board this train who speak German like a Berliner. A German of the militarist type and a Prussian, arrogance personified, is also on. I wonder who they are. . . . The more I see of militarists, the more I hate them. Murderers, the lot of them, brutes without conscience to whom lives mean nothing. I detest the Prussian type of militarist and Prussians in general." Rock did not realize that Chiang Kai-shek had been employing German military advisors since 1927 or that the top German military figure currently in China was no less than General Hans von Seeckt. But whether one knew who they were or not, the effect of all the guns and soldiers standing doll-like at attention was unsettling. After all, Rock had fled Europe in part to be far from warlike activities. He tried to still his anxieties with the thought that Yunnan would be quiet and removed.

In Shanghai Rock called on H. H. Kung, Chiang's brother-in-law, Minister of Finance, alumnus of Oberlin College, and one of the most powerful figures in Nationalist China. Unfortunately, Rock disclosed neither his reasons for, nor impressions of, the interview, only listing it matter-of-factly along with encounters of no special significance. At sea again, en route to Hong Kong, his freedom weighed heavily upon him and doubts plagued him. "Home! Oh, I want a home, but not a lonely one and yet I know there will never be another." It was just a mood and the worst of it passed in the night, though not without provoking some melancholic musings about time, space, existence, and the reality of the soul. The closer he got to Hong Kong, where he had stashed all his belongings, the more imperative it became for him to reach a decision, and China as he had seen it outside Peking made him uneasy. He grappled with himself for a few days in Hong Kong and then wrote with conscious deliberateness: "I must now make up my mind. I will make the following decision: I will stay in Yunnanfu until the autumn and then will go to Peking and will take everything with me and there I will stay!!!! Stay!!!" By mid-June, 1934, he was back in Yunnanfu.

Yunnan really did not seem like home, filled with friendly faces, including dear old Watson, and Rock was glad to be there. Three Nakhi men from Likiang had come to meet him; his favorite, Li Ssu-chin, had begged off on account of illness. Rock knew he was lying but could not blame him for wishing to stay with his family. The reduced entourage suited Rock's needs and pocketbook perfectly. He found an agreeable house, unpacked, and settled quickly into work on a new project, which he called the "History and Geography of Southwest China," an enlarged version of the work he had done on the Nakhi kingdom. He drew the history from Chinese sources and relied largely on his personal observations for geography. Save for his untrustworthy internal plumbing, he felt quite contented in familiar Yunnanfu. By the time summer ended he had the sensation of having just arrived and was not very keen on moving to Peking. He put off the trip to the east coast until November and then spent most of his time unearthing historical documents instead of hunting for houses, leaving a somewhat irritated Vetch to wrestle alone with the Nakhi manuscript. Rock liked Peking very much, but it had an unpredictable future. Besides, he was attached to Yunnan, and the idea of home obsessed him. Yunnanfu was as close as he could get to that illusion. He returned there for Christmas. "As far as I am concerned," he declared on his fifty-first birthday, "I feel better here, or more at home, than anywhere else. Here I live cheaply and have a nice house, good faithful servants. What more does one wish?"

In January, 1935, Dr. Watson and his family packed up to return to England, leaving a downcast Rock behind. He felt in his bones—correctly, as it turned out—that they would return, but was so sad when they left that he impulsively wrote a check for his sister in Vienna, as though to fortify his notion of family. At the end of March he fired his cook, Yang Ki-ting, who had been with him longer than anyone else, on all the major expeditions and, as Rock remembered, had often concocted meals under almost impossible conditions in the wilderness. But of late Yang had fallen under the spell of "an unscrupulous woman" and spent Rock's household funds recklessly. Loath to lose such an excellent cook, Rock tried to reason with him, but Yang could not be coaxed away from his romance and Rock finally had to discharge him and train one of the other Nakhis to do the job. These small changes aside, Rock's life in Yunnanfu acquired an almost permanent quality.

During tea with Lord Walter Rothschild the year before in London, the conversation had drifted to Chinese politics and communism. Rock had made Lord Walter a beneficiary of his political wisdom by explaining reassuringly that in China, "the poorest of the masses will not be bolsheviks but will become bandits which they, of course, are." At that moment in history, practically no foreigner took the Chinese reds seriously. Chiang Kai-shek, however, did not share Western optimism. In April, 1933, though Japanese troops were at that very moment penetrating the Great Wall and entering Hopei Province, Chiang ordered the fourth of his bandit-suppression campaigns. Unfortunately for the Kuomintang, it ended in defeat like its predecessors and produced an unhealthy level of frustration in Nanking. The German advisors had their work cut out for them. The fifth campaign, begun in October, 1933, had apparent short-range success, but a year later, just when Kuomintang victory seemed inevitable, the communists began their Long March.* Ninety thousand communist troops of the First Front Army broke through Nationalist lines and, abandoning their base in the Kiangsi-Fukein region, started to march westward with their foes in hot pursuit.

During the early months of 1935 the communists were preoccupied with provincial and Nationalist forces in Kweichow, who obstructed their path across the Yangtze into Szechwan. At the end of April, therefore, the Reds shifted to the southwest and headed for Yunnan. Lung Yun was immediately alerted to contain them south of the Yangtze; meanwhile, the marchers advanced rapidly toward the capital and panic took over the city. Rock's diary for May 1, 1935, reads as follows: "Martial law is declared in [Yunnanfu] and after 7 p.m. no one is allowed to be on the streets; the latter are barricaded. The communists are at Tapanchiao, forty miles from the city. They captured several truckloads of ammunition sent to Chutsing for Col. Kung, who was to come via Chutsing. Instead he came via Iliang, and the Reds intercepted the ammunition. [Lin Piao's division was responsible for this coup.] Another blunder: airplanes arrived in Yunnanfu but they have no gasoline for them and on their arrival they sent to Hanoi for gas. This was held by the customs on the Chinese side and it has not yet arrived in Yunnanfu . . . Another incomprehensible muddle: a telegraph office received orders to send telegrams for Yunnanfu via a place to the east of Yunnanfu. They took the messages plus the money and sent them, instead of directly to Yunnanfu, to a place to the east, where they were to be redispatched. The communists took the region and the telegrams repose there awaiting their delivery when the Reds should have left.

---

* It is not the purpose here to describe the Long March, the outlines of which have been drawn by many authors. Dick Wilson's account, published in 1972, helpfully fills in many details of the historical flight. The impact of the Long March on the Chinese people was brought to my attention by a Yunnanese gentleman who now lives in the United States. His family had once been large landowners, and the communists had confiscated their property in 1949; the boy fled and has not had any correspondence with his relatives for almost twenty-five years for fear of causing them trouble. Politically he is resolutely anti-communist. Yet he said, with undisguised pride, "You know, the Long March passed through my village!"

"I think the situation in China is irremediable. There is only one word for it: Chinese, which is synonymous [with] hopeless.

"The latest news is the reds are 3,000 strong at Hapanchiao and another lot had taken Fumin to the north of Yunnanfu and they await their reinforcements. It remains to be seen whether the reds will attack Yunnanfu at all costs or whether they will simply move northwest across the Yangtze to Szechwan. If I were Lung I would let them go, and to hell with [the Kuomintang general] who betrayed him. He forced Lung's troops to cross the Yunnan border into Kweichow and even to east of Kweiyang when the reds threatened him. . . . His own troops remained inactive and fresh as a daisy. His own troops did no fighting; he let the others do it [in] Yunnan, Szechwan, and Kweichow. The latter were decimated by the reds, and thereupon Chiang took control of Kweichow and he will do the same with Yunnan. Poor Yunnan, official ridden, squeezed, is made the scapegoat for that wretched militarist. It does matter apparently nothing if a few thousand people are impoverished and slaughtered as long as he gains what he is out for. From the tactics displayed it looks as if he were in league with the reds . . . He is forcing the communists to the south into Yunnan, closing their way in the north, but Lung undoubtedly will play him the trick and let them slip west . . . and Chiang will be in the lurch and it would serve him right."

Chiang and Madame, who had arrived in Yunnanfu to coordinate the campaign, "hastily repaired down the French railway toward Indo-China,"[1] in the words of Edgar Snow, as Lin Piao's division rushed toward the provincial capital. The communists were less than ten miles from the city walls when it was discovered that the main column had bypassed Yunnanfu and proceeded west, and that the threatened assault on the city was merely a diversive action. The communist leaders had thought like Rock. So, apparently, had Lung Yun, and there is evidence that Lung made some kind of deal with the communists. By May 9 the latter had managed—despite elaborate precautions by the Nationalists—to cross the Yangtze into Szechwan. It has been suggested that Lung's troops did not pose any "serious harassment" to the reds but, in fact, deliberately permitted them to slip across the river. Lung's position in the communist government after 1949, as vice chairman of the National Defense Council and vice chairman of the Southwest Administrative Committee of the Chinese People's Republic, reinforces this theory. The Lung Yun of 1935, however, was a hard-core warlord who had not undergone any ideological transformations. Whatever succor he had given the Red Army was motivated by self-interest. He wanted Chiang off his back, and the most obvious way to get rid of him was to make sure that the Kuomintang would be chasing communists in some other province.

During the few days when it appeared that the communists might sack Yunnanfu, Rock had taken the precaution to withdraw to Kaiyuan (Amichou), a city about one hundred miles to the south along the French railway. He had left the two Nakhis to pack and await further instructions. Kaiyuan was mobbed with refugees—foreigners and Chinese women with children who could pay the price of the train ticket—for about a week until it was clear that the capital was out of danger. Rock returned, and the Nakhis unpacked. The panic subsided. Life in Yunnanfu resumed its old patterns very quickly. The reds proceeded into Lolo territory in Szechwan, where their strategy was to enlist the sympathy of the barbarians against the common enemy, the "white" Chinese. Generations of ill will cannot be reversed in a couple of weeks, and the Lolos understandably mistrusted these unsolicited advances. In some cases, however, the Reds actually won Lolo cooperation and even a few recruits to their ranks.[2]

The First Front Army, with Mao at its head, had traversed Yunnan in less than two weeks and had escaped its pursuers without a major battle. So swift was its passage that its impact upon either the Yunnanese or provincial politics was negligible. However, it was remarked by those who actually had encountered the marchers that the Red Army conducted itself quite differently from most soldiers. The communist troops were well disciplined and, though poorly dressed, exhausted, and hard pressed after the Kweichow campaign, had neither burned nor looted but paid for their food or received gifts from the peasants. Some of the voluntary offerings of supplies were genuine expressions of political sympathy; more often they were gestures of self-protection extended by people accustomed to marauding armies and hopeful that their paltry gifts would prevent total destruction. That these soldiers were not thieves or murderers like others was a matter of great surprise. The communists observed this code of hospitality with remarkably few lapses throughout the duration of the March, thereby impressing many people. Even as dedicated an opponent of communism as Rock remarked that "the reds pay for everything . . . when they found it necessary to help themselves to potatoes [they] put a stake in the ground where they had taken the potatoes and tied the money in a little bag to the stake."[3] For the first time he detected subtle differences between these communists and the Feng Yu-hsiang communists he had known in Kansu.

Kuomintang soldiers, meanwhile, had made a large-scale appearance in Yunnan. By contrast with the Red Army, they acted precisely the way Chinese soldiers usually acted: they "steal everything and give the people a beating in addition," as Rock put it succinctly. They were badly paid, fed, and—German advisors notwithstanding—disciplined; they behaved like parasites because they were permitted to do so and because they were hungry. Consequently, they were unloved by those upon whom they preyed. As far as Rock was concerned, they were no better than ordinary bandits.

The survivors of the First Front Army—roughly 5,000 of the original 90,000 who had left Kiangsi—arrived in Yenan in October, 1935, after almost precisely a year on the road, and this is what is generally referred to as the Long March. There were, however, three other red armies which eventually joined the First in Shensi. One of them, the Second Front Army, was dislodged from its base in Sangchih, Hupeh, in November, 1935, and, like the original marchers, headed west looking for a place to cross the Yangtze. Their route to Shensi was as long, and their progress as bedeviled, as that of the First Front Army, but their story is, for obvious reasons, less well known. The Second Front's most prominent and colorful leader was Ho Lung, a one-time Hunanese bandit chief who had not converted to communism until 1927. Edgar Snow, who met him in Yenan, devoted a chapter of *Red Star Over China* to him.

The 40,000 soldiers of the Second Front Army began their trek by pushing almost due south into Hunan with the object, as Wilson explains, of collecting funds and dodging Kuomintang.

From there they veered west but made no advances toward the Yangtze. Like their predecessors, they reached Yunnan in April and once again threatened Yunnanfu. By the second week of April the capital had the jitters. Machine guns had been mounted on street corners and on the bridge and, as Rock said, "things did look exciting and disquieting."[4] The foreigners flocked to the railway station, and women already packed the train. Rock went home, paid off the Nakhis, packed his suitcase, and returned to the station to wait, but by dawn nothing had happened so he went back to the house again. All the following day, April 11, the day before Easter, he stayed indoors with the servants and listened to the roar of cannon and the explosions of bombs dropped from Nationalist planes. It was all very hard on the nerves.

The communists had in their midst a pair of hostages, two foreign missionaries. One of them, a Swiss named Alfred Bosshardt, had been captured by them a year and a half earlier and sentenced to eighteen months imprisonment for spying—"probably," as Snow noted, "no more than passing on information about red movements to the Kuomintang authorities, a practice of many missionaries."[5] As his full sentence had not been served when the communists left Hunan, he had been taken along on the march. They released him, with ironic symbolism, on Easter Sunday, and he had to be carried to the hospital in Yunnanfu. Rock hurried off to visit him out of curiosity and found him a complete "physical wreck," but not as bitter as one might have expected. "If the peasants knew what the communists were like," Bosshardt told Rock, "none of them would run away." Neither Wilson, Snow, nor Rock indicates the fate of the second missionary, Arnolis Hayman, which leads one to surmise that he did not survive his ordeal.

Though they treated the peasants well, the Second Front Army dealt harshly with officials and rich men. They disemboweled the magistrate of one town and skinned another alive; others were tied to tables and brought to the town square where they were chopped to pieces with an axe.[6] Fearful of what lay in store for them, officials and gentry in the marchers' path often tried to run away or hide. Red Army methods, if not delicate, were purposeful. "When they take a town," Rock explained, "the rich must pay. All they get is a receipt. The stores are bled and the goods distributed among the poor, also rice and other cereals from landlords who have more than they can eat. [The communists] have a band and Marxist literature and anti-religious slogans. Wherever they go they open a fair and distribute the things that belonged to the haves to the have-nots and so they are quite popular, for the latter class is in the majority. The rich are taken along and held for ransom." Sometimes, too, a few communists were left behind to organize the peasants for future political activities.

The Second Front Army bypassed Yunnanfu but continued much farther west than the First Front Army before they crossed the Yangtze. On April 25th they took Likiang—having made the journey at least five days faster than Rock usually accomplished it by caravan—and crossed the river even farther west. Rock heard a rumor that they had managed to take Muli and were trying to organize a soviet in the Batang-Muli-Likiang region. They negotiated the difficult mountain trails of Sikang, many of which were familiar to Rock, and dismayed their pursuers. In Kantzu, Sikang, they combined with the Fourth Front Army and went on to Shensi, which they reached in October.

Once the reds were out of Yunnan, Kuomintang soldiers moved in, looking remarkably fresh to Rock. "*Chinoiserie.* That is what the French call it. The Yunnan troops have stood the brunt of the fighting." But once more Lung Yun did not appear to have put up an energetic resistance. The Nationalists were "very unpopular, as they are nothing but bandits," Rock wrote a friend in Washington. "They loot everywhere, so that the people really wish the reds would come back. (Myself, however, not included!)"[7]

The net effect of the Long March Phase Two on Rock was to cause him to consider returning to Honolulu, where he could work in peace. Rock was sufficiently concerned during both communist incursions to send his library by train to Indo-China; when the communists moved on, he ordered the books back. "Here," he observed with uncharacteristic restraint, "one is always disturbed." "It is really the breaking of China with the Japs in the northeast."

In between Long Marches, Rock's studies progressed well.

Jaded by former upheavals, he did not worry much about Chinese politics. His life became almost routine, neatly divided between ethnological investigations and a rather gossipy social life with fellow foreigners—missionaries or government officials—punctuated by bouts of ill health and visits to the hospital. Complacency was not Rock's nature and he passed much of his time wondering if life might not be better somewhere else. Europe being in an unsettled state, he thought affectionately of Hawaii and wrote President David Crawford of the University of Hawaii that he was planning to leave the Orient for good and settle in Honolulu; the letter included an offer to loan his by now substantial library of Orientalia to the University. In fact, however, he made only one short trip, in late November, 1935, to Hong Kong. Otherwise he was quite sedentary.

Rock's big adventure during the year was his first flight between Yunnanfu and Likiang, which he made in February, 1936. He had not seen Likiang since April, 1932, when he departed in a foul temper because of the anti-foreign insults heaped upon him; but he had forgotten the unpleasantness, and the town occupied a sentimental corner of his mind. He had wanted to return many times but was in no physical condition to undertake the caravan journey. Eventually he had the idea to charter a plane, and early on a clear Monday morning he made his way to the airfield. The two Nakhis went with him so they could visit their families.

It was not his first flight; he had flown with two other Nakhis over Washington once before. Yet he immediately sensed the excitement of being airborne—an experience commonplace today. For Rock, also, there was the wonder of seeing in large scale the terrain he had so painstakingly explored and mapped from ground level, of recognizing familiar mountains or river gorges from an entirely different perspective. It was rather like suddenly meeting an old friend, whom one has always seen in overalls, wearing a tuxedo. As the plane approached the Likiang snow range, the flying became hazardous but always breathtakingly beautiful for Rock who, happily ignorant of the laws of aerodynamics, had little thought of danger. They landed, finally, in a meadow above Likiang. Rock had sent word ahead of his arrival to his old retainers, who had been watching the sky. Li

Ssu-chin, now leading the life of a yak herder, was the first to arrive (". . . he wore the suit I once bought from Hagen and gave him. He had changed a great deal. He looked yellow and haggard, very different from the days when he worked for me"). All of Nguluko followed to look at the plane rather than its passengers.

Rock, meanwhile, made his pilgrimage to the emptied village. "I went to my old house. I could hardly believe that I was back in the place I had spent so many years. . . . The wind howled as usually, the fields were still grey although the peas and wheat had come up. It was cold, yet the sun shone brilliantly and the familiar peaks stood out marvelously clear against a deep rich blue sky. I lived over the years of my sojourn during the two hours of our stay. Yet I would not liked to have remained behind this time. I ate my lunch sitting on the stone flags where Lau Ru used to sit skinning the birds. The screen had gone; a pig jumped through a hole in the door which led to our kitchen. . . . We left and I peeped into the schoolhouse where I had planted some trees, *eucalyptus* from California. I forgot to look for the apple tree I had planted in the little garden outside our house." All this he recorded in his diary on February 3, 1936.

By the time the party returned to the meadow two hours later, a mob—Rock called it thousands, probably an exaggeration—had gathered around the aircraft and people pressed their noses against the windows. Some of them had come to see Rock and he recognized many faces in the crowd, usually of people he had doctored years ago ("the old woman I operated on on a gangrened foot, the man who had been stabbed and whom I healed . . ."). Partings were, as ever, difficult; the brevity of the stay did not make things easier, as one might have supposed. Time was completely out of joint. By 4 o'clock the same afternoon, Rock had returned to Yunnanfu, having accomplished in eight hours what would have taken him a month and a half by the old methods. He made a long entry in his diary, as if to inject some reality into what had been a dreamlike day. He went back to work the next day as though nothing had happened.

Ho Lung and the Second Front Army came through a couple of months later and disappeared quickly. Soon after, Dr.

Watson returned to Yunnanfu, much to Rock's delight. One day at the end of August 1936, Yang Ki-ting, the wayward cook, reappeared on Rock's doorstep begging forgiveness. Rock, Mr. Well-I-Told-You-So incarnate, remarked that he looked like a tramp but "he knows better now. He himself said he feels as if he had been raised from the dead, a provident chap," and welcomed him back to the fold, a prodigal son returned.

Autumn found Rock on what was now becoming a ritual trip to the east coast to scavenge in bookstores, consult with Vetch, and ease the restlessness in his bones. This time, however, he thought of it strictly as a vacation and meant to return to Yunnanfu. Keeping in step with changing times, he made his journey by air. He flew first to Chengtu in Szechwan and was astonished to see how modern that city had become since his last visit in 1927. The balance of the trip was Nanking, Peking, and Shanghai. He spent six weeks in Peking, lavishly installed in the Hotel du Nord. There, by accident, he met Lester and Loy Marks of Honolulu. Vague acquaintance developed into what passed with Rock for real friendship. Loy Marks, a dynamic, brainy woman with enough wit and curiosity to keep him amused, was especially good company, and he enjoyed meeting with the couple for dinners in Chinese restaurants. There were also people he knew from way back, such as Edgar Snow, who asked him about the communists and to whom he related the saga of the missionary Bosshardt. Though not an eventful trip, it was academically useful and rescued him from provincial stagnation. He went back home in December and immediately contracted bronchitis.

By the time he recovered from his respiratory infection, his digestive apparatus malfunctioned again, as though there were some kind of conspiracy against his body. The London doctors had suggested an operation to improve his condition and, as he lay in bed on his birthday, sick as the proverbial dog, he toyed with the idea of returning to Peking and having it done. Like many people, however, Rock was afraid of surgery—both of the scalpel and of the temporary limitations such treatment would necessarily place upon his activities. As soon as he felt a little bit better, he stopped thinking about drastic remedies. But ill health made him susceptible to specters of sudden death or political catastrophe and impulsively, at the beginning of February, 1937, he packed up his belongings, including his by now precious library, and headed for Hanoi. He had Ho Chi-hui, his Nakhi scribe, with him.

Having planned no itinerary, much less a destination, Rock moved instinctively from one place to another, south from Hanoi by motor car to Saigon, through villages and cities whose strange names are now painfully familiar to Americans. Then he went to Angkor to see the wats again. If his diaries are an accurate indication of his reactions, the high point of the trip was Charles Chaplin's *Modern Times,* which he saw in Hanoi; tribute to Chaplin's genius is that Ho Chi-hui laughed as hard as Rock. Then indecision and back to Yunnanfu, but only briefly and suddenly, in April, to Hanoi again. Rock began to weary of his aimless peregrinations: "I wonder where I will end up, in Honolulu or Peking . . . I won't go to Europe, not under the present disturbed conditions, and I don't care if I never see that country [sic] again. I should learn once and for all that the best thing I can do is to stay in Yunnan in one place and await the end. I have a nice house and have faithful servants and good wholesome home cooking. I shall bring a radio and refrigerator and then I will have all the news and music and ice cream I shall want. Perhaps I shall send for my nephew to come and work with me. We shall see. I don't want strangers about me. . . . I shall see what I will do in Peking. Perhaps I will let them operate on me. All I care is to be left alone and in peace. Let society be society with its ranking business and jealousies and superficial[ities] and nonsense. I will live my own life in peace and do work to my liking, collect valuable books and then I shall be ready to drop this mortal coil."

So he went to Peking but, feeling very fit, did not allow an operation even though he knew he ought to; instead he enjoyed himself, saw Vetch, did his research work at the Palace Library and the National Library of Peking, and socialized agreeably. After a month he went on to Shanghai where he managed to make some publishing arrangements for his new manuscript, now nearly completed, and taking the Hong Kong-Hanoi route, was back in Yunnanfu at the end of June. He had his radio but, alas, no refrigerator, so no ice cream.

It appears that Rock could have stayed indefinitely with home base at Yunnanfu, making occasional journeys in whatever direction captured his fancy at the moment. He felt comfortable and familiar in the Yunnanese capital and no longer complained about the anti-foreign insults. Certain things, like the public facilities, never ceased to appall him. "On the main street," he could remember years later, "in the center of the road were a few stands where passersby could, for a few coppers, attend to their toilet in public. On a wooden frame above the table hung three or four public toothbrushes, and about two towels which were suspended from the covers of the frame. On the table were three washbasins . . . That these washstands with their towels and toothbrushes spread disease needs no emphasis." But he had seen much worse sights and, while no less sensitive to them, had taught himself not to flinch. Beggars, children smoking opium on the streets, filth, and disease could be avoided if one knew where not to go, overlooked if one simply had to pass them. Rock stayed out of their way as much as possible, circulating among the other Westerners. He was sometimes considered a bit peculiar, but always entertaining. He had become a foreign fixture in Yunnanfu, almost a dignitary, much invited to important functions. The role pleased him. He was often painfully lonely, as his thoughts about Robert indicated, but less so there than in other corners of the world. His utterances about leaving China forever were really quite empty.

All the same, China being China, it was foolhardy to think of Yunnanfu as a permanent residence. One always had to be prepared to leave at a moment's notice. Domestic politics were hopelessly tangled, and the Japanese were growing more aggressive every day. Now that he had a radio—not even the American Consulate had one; Rock used to copy down the news and deliver it to the Consul—he could listen to the BBC as the world deteriorated into war. Whatever passing admiration he had conceived for the Japanese converted itself into hatred: "The Japs are raising the devil in north China; some catastrophe should overtake these rats and bring them to their senses. Why such pests, imperialists of the first order, should be allowed to exist! They should be given a lesson and the haughty wretches should be taught to respect other people's rights. Yet the Chinese are just as haughty and overbearing and impudent, thinking themselves so much better than everybody else. Pride cometh before the fall." The most recent source of Rock's grudge against the Chinese was that he had invested heavily in Chinese currency following the Nationalists' economic reforms. He persisted in despising Chiang Kai-shek but, as he observed sourly, "if this government disappears, my money is gone to them . . ." At the moment, neither his money nor the government looked very secure.

By mid-summer of 1937 China and Japan were embroiled in what history books charitably call the Undeclared War. Considering the outrageous provocations of the Japanese since the Mukden Incident, it was a wonder that war had not been declared long ago, but Chiang had chosen to chase communists while limiting his opposition to the foreign intruders to vitriolic rhetoric. The Japanese had been permitted to nibble away in the northeastern provinces, and Tuchman describes the scene in Peking in 1935–36 as "Japanese military pervading north China, strutting the streets, knocking Chinese out of their way with blows of their rifle butts, dictating to puppet governors and officials, summoning press conferences and issuing fire-eating statements about Japan's 'divine mission' to lead the peoples of Asia."[8] Lack of resistance to military ventures had effectively muffled the protests of the anti-war faction in Japan; in response to international censure, Japan casually withdrew from the League of Nations, further crippling an organization that had been castrated at birth. Meanwhile, the Chinese—particularly those who had to put up with the insulting Japanese presence—became increasingly impatient.

Then, in December 1936, the famous Sian kidnapping of Chiang by his own forces abruptly altered Chinese policy. With the communists concentrated in Shensi, Chiang had determined to wipe them out once and for all, but the campaign had gone badly. Hoping to raise the morale of the troops and change the course of the fighting, Chiang flew north to Sian, the northern military headquarters. There he was taken captive by Chang Hsueh-ling, and the 105th division of the Northeastern Army mutinied. They had been infiltrated by communist propagandists and simultaneously discouraged by the lack of resistance to Japanese aggression. Chiang's champions quickly sent airplanes to Sian as a demonstration of their support, and China seemed precariously close to a renewed, large-scale civil war, a condition it could ill afford at a time when the Japanese foot was already in the door.

Suddenly, an improbable *deus ex machina,* the suave Chou En-lai, stepped forward as a mediator. The arbitration resulted in a verbal truce between communists and Nationalists, a united front against the Japanese. The agreement temporarily served the best interests of all Chinese factions and of the communists' backers in Moscow who also feared Japan and recognized that a unified China would relieve Soviet anxieties over its Pacific territories. The Sian bargain was absolutely cynical. Chiang never for a moment lost his obsessive hatred of the communists; the communists never abandoned their ideological objectives. Yet, when Chiang returned to Nanking on Christmas Day, the Chinese were free to fight the invader instead of each other.

Real warfare did not break out until more than half a year later, in July 1937. It quickly became apparent that, politically unified or not, the Chinese were no match for the Japanese modern military machine. Peking was evacuated quickly at the end of the month to save its cultural treasures from bombardment; the Japanese took Tientsin two days later. In August Japan opened a new front at Shanghai.

Rock, deep in the interior, did not budge. He sensed no danger, only the abstract rage that possesses people who oppose war in principle and from a great distance. His financial future did not look bright, his health had been mediocre all summer, and he was miserably lonely. On November 3, 1937, he wrote his nephew that he would make arrangements for him to travel to Yunnan. His need for family was so deep that he had managed to block out his memories of Robert's shortcomings. Robert had again been transformed by his romantic alchemy into a surrogate son.

XI

A SCIENTIST'S DILEMMA AS THE JAPANESE BOMBS FALL

There was something almost perverse about the pattern of Rock's travels during 1937. He was out of China during the first half of the year when it was quite safe to be in; and he insisted on staying—and plotting for his nephew to join him—during the second half when American representatives had been advised by Washington to recommend the evacuation of U.S. citizens from China. In the early stages of the war the hostilities only endangered Americans living in coastal areas, and consuls in the interior did not press the case for exit very vigorously. Nevertheless, by November 4, 631 Americans had quit China, nearly half of them from Shanghai and only 361 from the Yangtze River interior, leaving behind 5757.[1] Those who stayed were stubborn: there were still more than 2300 in Shanghai, and more than 1100 in the interior of which one was our protagonist, with his newly installed radio, feathering his nest as though he expected to spend the rest of his life in Yunnanfu. Scientifically exacting and thorough in scholarship, he was never much of a realist about his own life.

The music which Rock loved so passionately and for which he had bought the radio in the first place was seldom what he listened for as 1937 ground to its unhappy end. Two days before Nanking fell to the Japanese he heard discouraging reports via London. Streams of wounded Chinese soldiers fleeing the efficient Japanese war machine roamed the capital's streets in a daze, and no foreign observer had any hope that the Chinese could hold the city. The infamous aftermath, the so-called Rape of Nanking, during which Japanese soldiers massacred and molested countless civilians, was too nightmarish to grasp clearly. Chiang Kai-shek, meanwhile, announced the removal of government offices first to Hankow and then to Chungking in Szechwan, taunting the Japanese to pursue him into China's depths and exposing Rock's northern flank, so to speak. The freshly acquired strategic significance of the western provinces, especially Szechwan and Yunnan, made Rock, who had trusted in their remoteness, uncomfortable.

He was plagued by doubt and toothaches. He had two rotten teeth extracted and ordered everything packed. Then, though war bulletins continued to be alarming, but possibly because his dental problems had been solved, he ordered everything unpacked again. No doubt his faithful Nakhis, who had performed this ritual many times, wished he would make up his mind, but he lit upon decisions as a bee upon flowers, flitting from one to another. On Christmas Eve he sat glumly beside his radio listening to "all but good news, only murder and war," and some old Christmas carols in German. New Year's Eve was even worse. Ho Chi-hui showed up during the day, and his master's fastidious nose detected the scent of opium in his clothing; Rock felt personally betrayed, though the man swore he had not smoked. Moreover, surveying the last year, Rock had to admit that he had not accomplished much. The new Muli King was in Yunnanfu grudgingly delivering his tribute to Lung Yun, and Rock had arranged for him to ride in a motor car and baked a cake, but this was a brief diversion. As midnight approached he sat alone, a little soothed by the strains of the overture to *Don Giovanni* but writing, finally, "I am unspeakably lonely."

The new year did not begin any better than the old had ended, and Rock observed the grim situation by spending much of January in bed reading Churchill. "Will there be a world war?" he wondered in his diary on his 54th birthday. "It seems so absurd to bomb innocent people, ruin their homes, mangle their bodies, and spend millions doing it in order, as the fools say, 'to bring *peace* to Asia,' and to make them love the aggressor." The Japanese had pushed 500 miles into north China and, despite some costly Chinese heroics, seemed unstoppable. The atmosphere did not favor scholarly research and Rock was distracted by concern for his valuables. He sent out trunks of especially precious possessions for storage in Hong Kong: brass idol, old books, tsamba bowls, a tiger skin rug and the Tibetan costume in which he had posed in Choni. He stopped making regular entries in his diaries but listed the things he owned with compulsive precision. His academic achievements scored a resounding zero.

The unproductive life made Rock lonely and nervous and by mid-May he could not stand it for another minute. He left Yunnanfu abruptly on the 19th and started on a trip to Europe, leaving behind trunks packed with most of his library, records, diaries, photo albums, letters, manuscripts, maps, jade screens, brocades, photographic negatives, food supplies, and his radio with instructions that, should the Japanese reach the city, all was to be shipped to Hanoi. Circumstances permitting, however, he was determined to return.

First stop after Hanoi was Bangkok where, for sentimental reasons, Rock called on the American minister. It was almost like seeing the inept George reincarnated. "The building has been improved but not the minister," Rock quipped, "who came out barefoot to greet me, a regular bore, uncultured and dull as they make them. He had on a Japanese kimono and nothing else. . . . He mumbled a greeting which I did not understand. I was sorry I came. . . . I think Hunt was an improvement over the present minister. Hunt, at least, was friendly. When I asked [the present minister] what is the situation here now he said 'there isn't any.' He told me that the Siamese are not pro-Japanese exactly but they were neutral, which I don't believe. Jap flags and Siamese flags were crossed over doorways because Jap warships were paying a visit." It would not have mattered what the minister had said; anyone who received callers in bare feet could not, by Rock's definition of good form, be believed.

After Bangkok, Rock went on to Europe by air, hopping

across Asia on what by our standards were short flights, with a dozen stops. He landed in Berlin, where he had instructed Robert to meet him. Having missed connections and incurred his uncle's wrath during their earlier encounter in Venice, Robert wanted no mistakes this time; he arrived early and checked into the hotel where Rock had made his reservation. This turned out to have been a mistaken move. When Rock found Robert settled in the hotel—one of the first class expensive category that Rock frequented when traveling, he berated his nephew for having no respect for money and ordered him to cheaper lodgings while he, himself, stayed on in style. And so the courtship began anew but, since the personalities were fixed, scenes played in 1933 repeated themselves. There was only one real difference: Robert had taught himself passable English. Otherwise, during the day Rock bullied him and Robert, a bit of a whiner, permitted himself to be bullied; then, because Rock could not sleep with a bad temper, he apologized in the evenings. Rock dragged Robert around as he attended to his scholarly business. They went to the botanic garden where Rock pointed out the hundreds of plants bearing the small wooden label reading: J. F. Rock, China. He was also trying to arrange for the publication of one of his Nakhi articles in a German periodical which specialized in Oriental studies. The transaction probably would have been concluded had not Rock and the German professor with whom he conferred accidentally missed each other for lunch in a Berlin restaurant. Rock mistakenly believed he had been deliberately slighted, and though the professor went to extraordinary lengths to prove that there had been an innocent error, he stood firm in his indignation and demanded his manuscript back from the editors.

In short, Rock was not at his best during his visit to Berlin. He had come, in part, out of curiosity to see the Nazis with his own eyes. He had not given Hitler much thought since his English hospital bed. In fact, the notion that someone might put the Austrians to work rather pleased his energetic, orderly soul; he had always considered the people of his native land somewhat lazy and frivolous. But Berlin shocked him. He was not exposed to military parades and heard no speeches; instead, as sometimes happens, a trivial incident impressed him deeply. One day he saw a policeman apprehend a careless pedestrian, shake the man by the collar, and drag him down on the sidewalk like a dog. It seemed incredible to Rock that so simple an offense should provoke so brutal a response, and yet the other bystanders did not protest the scene. He noted afterwards that Berlin pedestrians were very disciplined, crossing only at intersections and only at ordained times. The scholars with whom he discussed politics spoke disparagingly of Hitler—including the professor who wore a Nazi swastika on his jacket. Then, too, the air was heavy with militarism. Even as violent a critic of Western civilization as Rock was shocked by what he saw in Fascist Germany yet, like so many others, he did not comprehend the utter seriousness of Hitler's intentions.

Politically speaking, Zurich came as a relief after Berlin and Leipzig, but Rock was still in a huff over the missed restaurant rendezvous. Scarcely a day went by without some kind of tantrum and, by the time Rock and his nephew reached Switzerland, Robert was ready to go back home to Vienna. This time, however, Rock had not made any firm decisions about whether he wanted Robert in China, and if the boy went back to Vienna it was possible that, owing to political conditions, he might not be allowed to leave again. To buy himself some extra time, Rock offered Robert enough money to live modestly in Milan for a few months and said he would send for him later, depending on what happened in China. Robert accepted this proposition; despite all the abuse he had taken, his desire for travel and his uncle's magnetism being stronger than his common sense.

They separated. Robert went to Milan and Rock to Paris, where he did some scholarly business and, for comic relief, visited the tomb of Marie Antoinette and mused over those who had lost their heads during the French Revolution. He half intended to go on to the United States but was feeling travelworn and decided to return to Yunnanfu. He had been absent only two months during which the Japanese had made headway in Yangtze valley, but not enough to scare him away from his adopted home. By August he was sufficiently lonely to have dismissed Robert's undesirable qualities. He sent money for passage to Hong Kong. Robert, also endowed with an active fantasy, left

Milan as a third class passenger on a Japanese steamer in an adventurous spirit, believing that once he settled down to working as his uncle's secretary, everything would be all right between them.

Rock stationed himself next to his radio listening to the sounds of 1938 which, if anything, were more depressing than those of 1937. Yet, like many people, he did not always know how to interpret what he heard. He recorded Hitler's blood and thunder speech at the Nuremburg party rally in September simply as "provocative and challenging." For one so inclined to overstatement as Rock, such restraint indicated uncertainty regarding Europe. In China, however, he had no doubts. The Japanese suddenly hit Yunnanfu at the end of September. Air raid sirens startled Rock from his breakfast at 9 o'clock on the morning of September 28: "The sirens screamed louder and all at once Ho Chi-hui and I, looking out from our sitting room window to the northwest, . . . heard a sickening thud and then [saw] a huge column . . . of earth and dust rise in the sky and hover there. I thought at first it was an incendiary bomb. Next minute nine Jap planes flew over our house, one having been hit; he gave out a streak of white smoke. They then flew toward the airfield being pursued by one lone pursuit plane, a student flyer by the name of Wong. The machine gun fire and anti-aircraft fire was intense. It made me sick at heart. I felt so helpless."

The city was defenseless. The Chinese had not prepared for a strike so deep in the interior and nothing had been done to protect Yunnanfu. There were no trenches or bomb shelters; machine guns which last had been mounted on the streets during the second Long March long had been removed; the only anti-aircraft weapons within the city were on an iron tower on the hill in the center, protecting Lung Yun's residence. After the first bomb fell, people ran helter-skelter in a panicked search for shelter. Rock heard that some had taken refuge beneath the willow trees near the west gate but had been mowed down by Japanese machine guns when the planes returned to strafe; the willow trees toppled like dandelions in the path of a lawn mower. Yunnanfu was thus initiated into the age of modern warfare. Surveying the wreckage later in the day Rock discovered that many had been killed, including women and children; the school had been damaged; and twenty patients at the British hospital had been wounded. He started packing.

The air raid sirens continued to shriek warnings for the next few days, but no Japanese planes came into view. Chiang Kai-shek arrived with his entourage to evaluate the damage, and Rock spent some time discussing affairs with W. H. Donald, the Australian journalist who had become Chiang's political advisor. On the first of October a half dozen foreigners, including Watson, Donald, and Meyer, the United States Consul, assembled at Rock's house to listen to the news. Everyone was very "jolly" as reports of the Munich agreement came over the airwaves; three bottles of champagne were consumed in celebration. Rock expressed the group's unwarranted optimism when he wrote in his diary later that night: "I hope the Japs will get it in the neck now!"

They still threatened Yunnanfu, however, and Rock did not dare stay there any longer. He packed his almost 4,000 volumes and immediately moved them to a sealed railway car at the railroad station, trusting that the Japanese would avoid hitting it out of fear of French retaliation. Two days after the champagne party he gave his Nakhis enough money to get back to Likiang and departed for Hanoi by air. The separation hurt him more than usual. The books followed him safely, and he arranged to store them temporarily in Haiphong.

Robert, meanwhile, had been in Hong Kong for nearly two weeks but had been unable to proceed farther because of the bombings in Yunnanfu. Rock had provided him with the address of Henry Corra, an old friend, and the Corras, taking pity on Robert's pennilessness and meaning well, had invited him to stay as their house guest. When Rock arrived on the scene he was beside himself that Robert had not gone to the YMCA rather than impose on friends. His nephew's apparent preference for leaning on others instead of fending for himself irritated him and the incident, coming on top of the hurried exit from Yunnanfu, put him in a diabolic mood. During the next few days he engaged Robert in a series of quarrels which ended with his announcement that the boy's English was, after all, insufficient. Without further ceremony he flew off to Honolulu, leaving Robert stranded in Hong Kong without the price of a return passage to Europe.*

Upset by the whole fiasco for which he knew he had only his misjudgment to blame, Rock started for Hawaii with some thought of sitting out the war in Honolulu and continuing his research there. The Hawaiian press, which viewed him as an authority on all things Chinese—a role he took no pains to deny—greeted him with questions on the war. The British, he claimed glibly, had not been alarmed by the fall of Canton to the Japanese. "Residents in Hong Kong," he was reported to have said, by the Honolulu *Star-Bulletin*,[2] "appeared to be more concerned over a possible shortage of vegetables," and quickly repaired to Matlock Avenue where Harold Lyons put him up in a guest room. Rock surveyed the possibilities of permanent residence; they were not encouraging, particularly in financial terms. The cost of living was high and, though he still had his savings, he could not support himself for very long in the 'style to which he had grown accustomed' on the islands. The alternative, of living very cheaply and restricting his travels, was unacceptable to him. The more he considered it, the less he liked the idea of staying in Hawaii.

As long as he was there, however, he determined to make the best of his time. He finally submitted to surgery on his intestines. It helped, but only partially. He also solved the problem of what to do with his library by entering into an arrangement with the University of Hawaii. The University agreed to appoint him Research Professor of the History, Geography, and Botany of China at a stipend of $3,000 per year with the understanding that Rock would donate his library to the institution, either as an outright gift before his death or in his will. The appointment in no way presumed Rock's presence in Honolulu; it was, rather, a

---

* When Robert asked him for the fare, Rock refused and also advised friends, such as the Corras, not to lend him anything. Robert did odd jobs, as a gas station attendant, teaching German, etc., but could not collect enough money to make his way home. He finally managed a loan from some German friends in August, 1939, by which time war had broken out in Europe. His parting communication was a note to Rock reading "Dear Uncle, Thanks for everything!"

way of purchasing his very valuable books on an instalment plan. Rock, meanwhile, assured himself of a small, steady income, quite adequate for life in Yunnanfu or wherever. When he returned to the Orient at the end of 1938 he arranged to have the books he had evacuated from Yunnanfu shipped to Hawaii for storage. He did not yet relinquish his title to them and considered them on loan to the University.

Having moved everything out of China, Rock was unwilling to move it all back until he could do so with reasonable certainty that he would be able to work in peace. The Japanese continued sporadic bombings of Yunnanfu, hitting once every two or three weeks. An air raid system warned the city an hour or two in advance, and people streamed out into the countryside. On the whole, the damage was minimal, but the constant possibility that one might accidently wind up as a bomb casualty was troubling to the imagination. Rock determined, therefore, to perch on some quiet spot near China and keep an eye on things. He chose Dalat, a city in the South Vietnamese highlands and then the summer capital of French Indochina, as his retreat and, from Hanoi, wired his Nakhi servants to join him there. They replied negatively, but Rock would not give up and went to Yunnanfu to issue his orders from closer range. He prowled around his old house, now gloomy and empty, and would have liked to take it back; it was later hit by a bomb. He spent Christmas at the American Consulate with the Meyers. His persistence with the Nakhis paid off, and one by one they arrived in Yunnanfu, until eventually he had a menage of four to accompany him to Dalat. "The climate is wonderful," he wrote to E. D. Merrill in February, in Dalat, "and the whole landscape beautiful and the place is clean, not like Yunnanfu." He rented a house with three bedrooms and bought a car.

That year in Indochina was quite a happy one for Rock. After two years of disorder—some of which had been his own creation—he managed to settle down to intense work. He returned to his studies of Nakhi languages and customs and made real progress. Two works were at the printers: some translations of religious ceremonies, financed by the Harvard-Yenching Institute, and his study of the Nakhi kingdom. Vetch was having the latter set in Shanghai where, because of the occupation, work on it was hopelessly stalled. (The printers' plates for the religious ceremonies, also set in Shanghai, were destroyed later in a bombing raid.) When he got an urge to interrupt his sedentary routine he went in search of birds for the U.S. National Museum, which he offered as donations. Though conscious of his age and ailments, he was still full of energy and did a respectable amount of hiking on these small expeditions and came to know large sections of Indochina quite well. He could not speak any of the native languages but enough French was spoken by both the colonizers and the colonized to prevent him from feeling like an outcast. Almost before he realized, 1939 ended, and he observed its passing on a heavenly peaceful night in Angkor. "I hoped," he wrote in his diary, "that 1939 would be a better year but alas, the world is mad. Were it not for Hitler and his bandit friend Stalin the world would be at peace. When will the Germans learn that war leads to misery?" The Japanese still occupied huge chunks of China, but there was little fighting. Yet Rock had remarkably few personal complaints and was contented enough to stay peacefully in Dalat for another six months.

Then, in July, 1940, the Japanese, greatly encouraged by Germany's success in France, determined that Indochina would make a perfect launching spot for their campaign in the South Pacific. The Japanese ambassador to France politely informed the Vichy government on July 17 that the Japanese army would begin its operations in Indochina on July 24. The gentlemen at Vichy had precious little room in which to maneuver and, on the day before the deadline, replied that they would not hinder Japanese entrance into the colony. These developments displeased Rock but might not have alarmed him had not the Japanese begun their peaceful occupation by dropping some bombs on Haiphong, presumably by mistake. Rock saw, suddenly, an uncertain future in Dalat; he also mistakenly assumed that the Japanese had seized Indochina as a base for attacks against Yunnan and southwest China. Japan had succeeded in isolating China from the rest of the world by pressuring the French to shut service on the Hanoi-Yunnanfu railway and on the British to close down the Burma road. Rock barely managed to get three of his four Nakhi servants back to Yunnanfu before the trains stopped running. Now, thinking that Japan would press farther into China, he wanted to get out of Asia. He left quickly for Bangkok with Ho Chi-hui, his remaining Nakhi, with the intention of taking him to Hawaii; he changed his mind, however, and packed the servant off on a flight to Yunnanfu while he boarded a boat to Manila and then a plane to Honolulu where, if necessary, he resigned himself to staying for the duration of the war.

He gave up this idea very soon. After his arrival, Rock went over to the University to see what had become of his library. Much to his horror he found the books lying carelessly in an old room under attack by cockroaches and bugs of various descriptions; some of the packing cases had not even been picked up from the waterfront warehouse. He approached President Crawford in a state of advanced agitation and told him emphatically that the books must be properly stored. "Dr. Crawford replied that was the way he chose to keep them," Rock later told newsmen,[3] "and if I didn't like it I could terminate my contract. . . . I can take the damage to my books, but never Dr. Crawford's insults." He resigned his university position and withdrew the offer to donate his library, privately congratulating himself on his good sense in not having given everything away two years ago. He had the books transferred to an insect-proof storage area and now offered the collection to the Harvard-Yenching Institute. As for Hawaii, he went sour on the place and announced his departure from the islands. Noted Savant Quits UH in Huff: Cites "Insults" read a headline in the Honolulu *Advertiser* for August 28, 1940.

Asked about his future plans by reporters, Rock replied "You may say I am leaving Honolulu for an indefinite stay in the Orient."[4] He flew to Hong Kong and on to Shanghai to arrange for a flight to Yunnanfu. W. L. Bond, executive vice-president of the China National Aviation Corporation (CNAC) remembered Rock appearing in his office looking "like a healthy and rather mischievous cherub," charming and spoiled as ever, and intent on retreating to Likiang. He went to Yunnanfu by airplane, to Tali by automobile, and north to Likiang ("too far away for any trouble") by sedan chair. He came equipped for a long stay.

269

THE CHINESE WESTWARD MIGRATION

The migration of the Chinese people westward during the Japanese occupation of the eastern provinces of their country was, in many ways, as heroic as the Long March. Nothing illustrates so dramatically the vastness and invulnerbility of China as the fact that millions of people could withdraw under the shelter of its unoccupied territory; and nothing demonstrates more clearly the stubborn courage and resilience of China's people. No one actually counted the number of Free Chinese who deserted Japanese-held areas and Wang Ching-wei's puppet regime in Nanking, but they came from all segments of the population: rich merchants, peasants, laborers, shopkeepers, students, intellectuals, camp followers, children, old people, whole families. Estimates of the number of persons who went inland range all the way from 3 million to 25 million, according to *Thunder Out of China* by Jacoby and White. Entire villages moved as a unit, their inhabitants burning everything they could not transport so as to leave nothing behind for the occupying forces. They brought whole factories and universities and rebuilt them in new locations. The north Chinese universities merged in Yunnanfu and remained there for the entire war as National Southwest University. "The students were camped four, six, and eight to a room, some of them domiciled in a rat-ridden, cobwebbed abandoned theater; they ate rice and vegetables and not enough of these;" wrote Theodore White and Annalee Jacoby, then staff correspondents for Time, Inc., who also said that "thousands crusted the junks moving through the gorges; hundreds of thousands strung out over the mountain roads like files of ants winding endlessly westward. There is no estimate of the number who died of disease, exposure, or hunger on the way."

The great migration ended by the summer of 1939. Those who had survived set about to resume their lives in Free China. Many of the refugees faced more than alien surroundings, strange terrain, disease, hunger and lack of shelter; Chungking, which had absorbed 800,000 new residents, was under air attack after May, 1939, whenever a friendly cloud cover did not protect the city from Japanese bombers. Air raid shelters were non-existent, but Chinese resourcefulness came to the rescue, an efficient air raid warning system was developed, and the caves in the rock cliffs below the city were transformed into sanctuaries. When the Japanese flew overhead discharging their grisly cargoes, the Chinese burrowed beneath the capital of Free China like moles. Still, the casualties were appalling.

The forced encounter between coastal and interior China had at least as great an impact on the nation as the Japanese occupation and military events. The removal of the Nationalist government to Chungking and the westward trek of the Free Chinese spelled the end of centuries of isolation for the interior provinces. A yak herder grew accustomed to the drone of airplane engines overhead, new items appeared in the market place, new roads were built, peasants saw trucks and jeeps for the first time. The terrible Tebbus of Choni, the unruly Ngoloks, the Lolos, and the Konkaling bandits mights still be law unto themselves, but with a cinema playing to packed audiences in Tatsienlu, a new order was in sight, for better or worse. As for the refugees, few of them knew anything about provinces like Yunnan, Szechwan, Sikang, or Kansu; fewer still had ever been there. The great mountains with their foaming streams, the climate, the diets—everything was different from the places they had left. Tibetan horsemen, Mongolian nomads, Lolos, Nakhis, Minchias, Lissus—the barbarians of history, were strange people wtih exotic languages and customs. No coastal Chinese had ever tasted yak butter, and now, tasting it, he did not like it. Chinese realized for the first time the magnitude and variety of their country; in this respect they shared a revelation with the Long Marchers.

On a practical level, because it was effectively cut off from the outside until Americans began flying some supplies over the Hump in 1942, Free China tried to fall back on its own resources, for the most part with pitiful results. The western provinces had been economically neglected for too long to respond quickly to the needs of a modern wartime army and a squatter population of millions. There was practically no industry; overland communications had barely improved since Marco Polo, much less since the days when Rock moved from one town to the next with mules and coolies. Yet, though far from being realized, the economic potential of inland China was recognized for the first time on a large scale.

Circumstances made it difficult for either the refugees or their reluctant hosts to indulge in philosophical speculation about the implications of this congregation; more ordinary matters compelled their attention. Many of the coastal Chinese, and especially the wealthier people, accustomed to the sophisticated environs of cities like Shanghai, Peking, or Canton, were shocked by the scarcity of physical comforts and dismayed by the lack of what they considered cultural refinements in the interior. They found the resident Chinese tiresomely provincial and backwards. The pressures for survival robbed the landscape of its glory and the tribespeople of their novelty and interest. Meanwhile, though the newcomers had introduced fascinating paraphernalia, they offended with their modern ways—women with lipstick and frizzy hair—and they had, by their great numbers, forced the prices of necessities sky high. The inability of the government to control spiraling costs produced a murderous inflation; that wartime phenomenon, the black market, flourished.

It was also observed in certain quarters that hardship did not necessarily bring out the best in individuals. Rock was especially bitter about the wheelings and dealings of a Chinese "botanist" who chose Likiang as his refuge. Soon after his arrival he appointed himself "forester" of the region, set up office in an old agricultural schoolhouse, and went into business selling permits for the cutting of lumber in forests where the Nakhi had been freely cutting timber for their homes for centuries. His organized band of armed ruffians collected squeeze money. Corruption, which had long occupied a prominent place on the Chinese scene, became more commonplace than ever, and because people like the delinquent botanist were not punished, appeared to receive official sanction.

There was, also, a war of sorts in progress. Those who lived in larger towns like Chungking and Yunnanfu could identify war with Japanese bombers, but for millions of Chinese peasants and tribesmen the war could only be felt in less direct ways. The commonplace notion that citizens rally 'round the flag in times of national crisis did not apply in China. Particularly in the western provinces where even provincial unity was unknown, generations of people had lived and died ignorant of the concept of national unity. The most that could be said during the war was that the people of China universally loathed the Japanese. To his misfortune, Chiang Kai-shek did not exercise the leadership that might have converted this hatred into a sense of national purpose.

Chiang's instrument of war, the Chinese Army, stirred more terror among the Chinese than it did among the enemy. Conditions within the army for non-officers were known to be so squalid that being drafted was considered only slightly preferable to death. Rock witnessed Chinese Army procedures in Likiang and offered the following account:

" . . . [only] sons were conscripted, while the sons of magistrates and generals were registered in schools which they never attended so that they could not be conscripted. Those who had silver could buy off the recruiting officer. The recruits in west China underwent no medical examination. The recruiting officer, usually a corporal, came to the Yamen, and the young men [also came], so many from each village. All depended upon the height of the men. Those who could afford it bribed the recruiting officer who declared that so-and-so was under the regulation height, and he got off. The particular village had to send another man. . . .

"Those so recruited were tied up with ropes, six men in a double line. Their arms were tied above the elbow and they also had ropes around their necks. I saw them assembled on the streets of Likiang, their parents, sisters, wives staying on the edge of the road. . . . When they wanted to go up to [the recruits] to say farewell or give them a message, they were kicked by the soldiers with their rifle butts. . . . Often when [the recruits] arrived at their destination they were told they were not wanted but were given nothing to find their way back again. . . . Those men had no money; they had to beg their way back, walking over high mountains without food, without adequate clothing, and many of them died on the road of starvation or from weakness. . . .

"In the Lung-wang Miao, or Dragon King Temple of Likiang, a small regiment was lodged for a few weeks and they had with them a young Chinese doctor who spoke English. He complained to me saying, 'I am absolutely useless here for I have nothing to work with, not a single drug or even a bandage.' He used to send his sick soldiers to me with a note saying 'Please give him such and such a drug.' It became so that he sent men with a prescription and wanted me to fill it as if I had a drugstore. . . .

"One day a group of soldiers wearing long gray cotton mantles with a red cross sewn on the left side, came to my house. They were invalid soldiers who had been sent west from central China; some were blind; others had limbs missing; they supported each other. I invited them into the court of the house where they sat down. I gave them tea and food and asked them whence they had come. Some were from Hupeh, Honan, or Hunan, and as they were cripples and useless, they were sent west. When I asked them what the Government or the Chinese Red Cross gave them they produced a government certificate to permit them to beg."[5]

Rock's observations of the Chinese Army echoed those of other Westerners who saw it at close range, and the list of accusations was long. Corruption was almost universal among officers who were, themselves, underpaid. They withheld food and money from recruits to pad their own salaries and could be bribed for almost any reason one could imagine. "The kindest thing to say about some leaders of the Chinese officer corps is that they were incompetent," said White and Jacoby.[6] "Besides thieving their men's food and money, ignoring their sickness, and flaying them mercilessly for infractions of discipline, they were bad leaders. Their staff work was inefficient. They reported to their superiors not what the situation actually was but what they thought their superiors wanted it to be." As a result of army abuses, the desertion rate was high and army morale so low as to be fictitious. In the absence of sufficient food and clothing, the soldiers stole from the farmers to stay alive. They stole desperately and without pity. As usual, the Chinese peasant bore the brunt of the war. The government taxed them, the Army recruited their sons and then robbed them. Those of them who had experienced the passage of the Long March made inevitable contrasts.

The blame for the wretched conditions within the Army lay with the traditional role of the military in China and with Chiang's failure to accept the introduction of reforms. The Chinese held military men in low esteem. For generations it had been customary for officers to depend on squeeze money; and, as

Rock had said any number of times, soldiers and bandits were often indistinguishable. What had been true during the warlord era remained true during the Second World War. Foreign experts, such as "Vinegar Joe" Stilwell, who worked directly with the Army, did not doubt the essential courage of the Chinese soldier or his ability to fight, given sound leadership. But the initiative for change had to originate at the top of the military structure and it was not forthcoming. The result was a demoralized and rapacious army of limited fighting capability and, equally serious, a demoralized and resentful people.

As long as Rock insisted on staying in China, Likiang was as good a place as any for him to be. The Japanese never came near the city; the closest bomb, Rock reckoned, dropped 150 miles away. He was, therefore, left in relative peace to work on his dictionary of the Nakhi language, a project on which he had labored intermittently for several years. He lived in a house in Likiang proper, rather than in neighboring Nguluko, and established himself in comfort. He and an Englishman shared the one radio in town, so the Chinese officials came to them for news. There was no problem of food shortage, but the luxury items that one used to be able to buy—coffee, tea, butter, cocoa, white flour, baking powder, sugar, meat—had either become outrageously expensive or disappeared entirely from the market. Having expected worse, Rock did not complain. He had brought along his own seeds, planted a vegetable garden, and lived happily on cauliflower, Brussels sprouts, beans, tomatoes, broccoli, lettuce, green peppers, spinach, and peas; occasionally he indulged in some meat. His principal source of anxiety was his health, always unreliable, but there was a reasonably good doctor now and a Cantonese dentist. There were times when his stomach reminded him that his impulsive flight from Honolulu had not been a rational decision. He was, of course, constitutionally restless. His injured dignity staged a recovery sometime during the summer of 1941, and Hawaii began to look good to him again. He sent a message to Arthur N. Young, Chiang's American financial advisor and director of the CNAC, indicating that he was anxious to return to the United States and asking if a plane could be sent to fetch him. The CNAC had far too many problems to cater to individual whims, and their immediate response was negative; however, a few months later they had to make a map survey in what would eventually be the Hump area and notified Rock that, weather permitting, a plane would land near Likiang to pick him up on the afternoon of November 23.

Rock waited at the airfield but, inexplicably, signals had become crossed. Instead of landing at the new official field near Likiang, the plane used the meadow on which Rock had landed on his first flight in 1936—about ten miles from Rock's waiting place. Bond, who was on the plane, noted that Rock's instructions had not been very precise. CNAC had promised to wait for fifteen minutes but, because Rock was nowhere in sight, they extended their stay to one hour. Rock, meanwhile, from his position in Likiang, had heard the hum of engines and guessed the situation. He commandeered every man and animal in sight to transport his luggage and dashed for the meadow. On the way he saw his plane take off. "To say he was frustrated would be a masterpiece of understatement," Bond remembered. Rock aired his wrath in a scathing letter to Young, accusing the latter of lacking all human virtues, notably honor and dependability. (When Bond reminded him of that letter years later, Rock laughed over it like a naughty child.) Rock was still stewing over his misfortune and wondering what to do next when he heard that the Japanese had attacked Pearl Harbor. He reconsidered his predicament in the light of this new information and realized that he might, for once, have been lucky.

His second thought was for his library, still sealed in an insect-free storage area in Hawaii, and for old friends. He wired Lyon in a panic to find out what had happened and learned happily that everything was safe. A few months later, when it seemed probable that the Japanese had no further designs on Hawaii, Rock had hopes of getting out to India via Yunnanfu. Had he been agreeable to leaving all his possessions behind, he might have managed an exit, but the war made transportation complicated and he could not dictate terms. He stayed on in Likiang where he considered himself "marooned" but, since his health did not fail him, he was marooned in reasonable circumstances and worked well. He punctuated his scholarly pursuits with attention to the crippled and ailing soldiers who appeared in his courtyard almost every day: he had earned the reputation of being a soft touch.

The longer the war dragged on, the harder life became for the peasants of Likiang. Though they were safe from the Japanese, they could not hide from the tax collectors. Chinese economy had gotten completely out of hand, and the government responded to the inflation with the simple minded expediency of printing more money. With the value of currency dropping by the hour, the peasants were absolutely unable to pay their taxes. The government, therefore, demanded payment of taxes in kind, of whatever grain was cultivated on a given area of land. Since the levy was based on land area rather than on percentage of the crop, people with fertile land were taxed less than those with poor land. The grain was measured by the *sheng*.

"The *sheng* was of a standard size in the district of Likiang," Rock explained, "but the magistrate who took charge of the grain brought into the Yamen made his own *sheng* . . . so that ten of his *sheng* made eleven standard *sheng*, thus taking one *sheng* squeeze for himself. Many of the peasants of the Likiang district brought their grain taxes six or seven days' journey from the outlying [communities]. This grain had to be carried on mule back, the mules had to be fed, they could perform no other work, and when they arrived at the Yamen they found out they were short so many *sheng*, for the magistrate used his larger measure. As these people carried no money, they were forced to sell an animal for whatever they could get on the spur of the moment and buy the grain to fill the magistrate's illegal measures; had they protested they would have been put in jail." Sometimes the magistrate would tell the peasants that the grain was supposed to be milled, and they would have to carry their grain back, husk it, and redeliver it to Likiang. The ultimate irony was that Likiang did not have any proper space for grain storage, so the magistrate had boarded up old temples and stashed the grain inside. It lay there awaiting collection which, because of the bad trails and worse administrative organization, was slow in coming. Insects and weevils feasted with delight, and Rock could see the resulting fine dust sifting out at the base of the boards. One year the provincial collector never arrived, and the magistrate ordered the peasants to pick up the infested grain from the temples and deliver a double amount the following year.

Rock had never felt greater compassion for the farmers, and he summed up their plight in a few words: "their money was worthless, their time wasted, and their grain spoiled. I need say no more. That they did not love their government was a certainty." That Lung Yun still controlled Yunnan was strictly an academic distinction. The Nationalists permitted his outrages, and the peasants did not distinguish one government from another. Whatever affection Rock had imagined for Chiang Kai-shek, out of sympathy for China in its fight against Japan, disappeared.

Japanese sights focussed to the south, leaving a suspensful China to await the outcome of their gambit in the Pacific. With the bulk of their military resources committed to this venture, the Japanese did not press for further gains in China. They were content to have isolated Free China from its prospective allies and to remind Chungking of their presence by intermittent bombing of interior cities. First the Soviet Union and then the United States was eager for the Chinese to engage the enemy but Chiang Kai-shek—the Generalissimo—skillfully stalled for time. By April, 1941, the Russians were involved in the European war theater and, in the absence of Chinese activity on the eastern front, signed a nonaggression pact with the Japanese. After Pearl Harbor, when the United States tried to lure him into battle, Chiang argued that the Chinese Nationalist Army was neither properly prepared nor adequately equipped to take on the invaders. Those who had seen the Army in action could not reasonably dispute his claim, but arguments raged over how to remedy the situation. The Generalissimo was much enamoured of Claire Chennault's theory that Japan could be defeated by massive air attacks initiated in Free China; General Stilwell, the chief American advisor to the Chinese and the U.S. commander of the China-Burma-India war theater, was an infantryman. In *Stilwell and the American Experience in China,* Barbara Tuchman supplies a complete account of he Chiang-Stilwell controversy and several historians have discussed Free China's wartime role, so it is not necessary to get involved in minutiae. Certain broad developments, however, deserve attention.

One observes first that the Generalissimo won his running battle with Stilwell. He disallowed or deliberately complicated the reforms that the American believed would have converted the Nationalist Army into an effective fighting instrument, and he obtained millions of dollars and military supplies under the

U.S. Lend Lease program, the United States (though not Stilwell) being distracted by more urgent problems and a little concerned that Chiang might make a separate peace with Japan. Much of the material that came so laboriously over the Hump after 1942 went to Chennault's 14th Air Force based at Yunnanfu; a lesser, but noteworthy, amount was hoarded by Chiang, presumably for use in some crisis. Secondly, relations between the Nationalists and communists deteriorated rapidly during the occupation. The uneasy United Front which grew out of the Sian Incident was in ruins by 1941. The communists, operating from their Yenan base, carried on a harassing guerilla campaign behind Japanese lines but did not neglect their domestic objectives. They succeeded in widening substantially their political authority in the northern provinces, thereby alarming Chiang who viewed them—and rightly so, as long as he refused to make any political accommodations with them—as equally dangerous to him as the Japanese. In 1939, in an effort to contain the communists, he began to throw a military cordon around the regions which they held; in 1941, when the red's New Fourth Army approached Shanghai, the Nationalists ordered it to withdraw across the Yangtze and then, while the communists reluctantly obeyed at considerable peril to themselves, attacked red headquarters. The "New Fourth Incident" dashed any outstanding hopes for a reconciliation between the factions; Free China was irrevocably split. Many historians and observers blamed Chiang's obsessive hatred of the communists for his failure to commit his army against Japan. By 1944 he had half a million soldiers facing communist positions in the north.

Rock, of course, knew neither more nor less of large-scale patterns than anyone else who listened to the radio, and he could easily be dislodged by rumors. He heard during the summer of 1942 that the Japanese might take Tali, which would have put them within easy striking distance of Likiang. He sped hastily to Yungning with the intention of continuing north to Muli and eventually to Tatsienlu; instead, he got sick in Yungning and stayed on the island of Nyorophu—somehow not the same without the old Tsungkuan. By the time he recovered, about ten days later, the Japanese threat had ended and he returned to Likiang grateful that he had not had to negotiate the mountain trails through Muli. He went back to his books and his sorcerer and his vegetable garden and his radio and the stream of ailing civilians and soldiers who begged for his attention. He remained in Likiang longer than he had ever stayed anywhere since Hawaii and, though he lived on the periphery of the war, even as a captive of it, he did not participate in it. Likiang's solitary military installation—if it could be called that—was a weather station to help American flyers get over the Hump. Rock, therefore, viewed the war's sad and peculiarly Chinese consequences from the sidelines.

Inevitably, owing to his health, disinclination to remain in any one place for very long, and the new offer of a post at the University of Hawaii, he often thought of leaving Asia. He succeeded, finally, in February, 1944, by traveling overland to Yunnanfu and taking a CNAC flight to Assam; from there he advanced to Calcutta. CNAC had by then established a regular service from Chungking, via Yunnanfu, to Calcutta, over the Hump. Uncertain how to proceed farther to his destination in Honolulu, he marched into CNAC headquarters, which had relocated from Shanghai, where he found a surprised W. L. Bond. Bond half expected an outburst over the 1941 Likiang plane incident but Rock, however temperamental, did not hold grudges. He was, on the contrary, in a cheerful mood. During their conversation, Bond had an idea.

The CNAC operated in cooperation with US Army Airforce officials in the Hump airlift. Given the capabilities of aircraft in those days, the Hump was one of the most frightening air routes imaginable. White and Jacoby wrote that it "drove men mad, killed them, sent them back to America wasted with tropical fevers and broken for the rest of their lives. Some of the boys called it the Skyway to Hell; it was certainly the most terrifying, barbarous aerial transport run in the world. . . . In some months the Hump command lost more planes and personnel than the combat outfit, the 14th Air Force, that it supplied." One of the problems was that the Hump never had been properly mapped, particularly the mountains. In order to avoid possible high peaks, the pilots flew at excessive altitudes where they encountered a lot of turbulence.

Bond knew of Rock's explorations in the mountains and realized that his familiarity with the terrain along the Tibetan borderland could be immensely useful to the Air Force. Rock probably knew more about the Chinese side of the Hump and the people who lived there than any American at that time. Bond took it upon himself to discuss Rock with American military authorities in Calcutta, who were immediately interested, and further introduced him to the proper officials. Almost before he knew it, Rock was hustled aboard a top priority flight to Washington. This unexpected mission suited him perfectly. He had an opportunity to make a substantial contribution to the war effort; he received a free ride to the United States, and he was treated like a V.I.P. which, under the circumstances, he was. His sense of duty, his pocketbook, and his vanity were thereby simultaneously satisfied.

Rock worked for the Army Map Service for a year, until the end of March, 1945, first as Expert Consultant and Geographic Specialist and later with the less glamorous title of Research Analyst. He performed the job with pleasure, with the knowledge of its value, never resenting the interruption of his own studies. All the time he plotted his return to China and observed with satisfaction as the military pendulum swung farther to the side of the Allies. The war ended in the Pacific, and China went back to normal. As 1945 ended Rock noted in the back of an old diary that "China is having civil war, as in the days twenty years ago when I wrote in this diary. I have just heard on the radio that Chiang Kai-shek is laying down the law to the communists. It will not be long before I start back to China, I presume. It will be my last trip, for I hope to remain and die there among my Nakhi friends."

That same New Year's Eve the Emperor Hirohito told his people to forget the legend that he had descended from the sun goddess and ended a chapter of Asian history. Things were changing in the Orient—and much faster than Rock ever dreamed.

XII

COLLAPSE OF THE NATIONAL REPUBLIC; COMING OF COMMUNIST CONTROL

Rock had excellent reasons for wanting to go back to China after the war—though one suspects that, had reasons not presented themselves, he would have invented them. When he flew to Washington from Calcutta in 1944 on the Army plane, he left his bulkier belongings to be sent after him. The ship that carried them to the United States intercepted a torpedo from a Japanese submarine and came to rest on the floor of the Arabian Sea. Sealed in Rock's trunks were translations of religious ceremonies, notes on Muli, and a rough manuscript for his dictionary of the Nakhi language—products of years of intermittent study. Coincidence and the Japanese Navy had conspired to wipe out a decade of research in one stroke; the irony was that the tragedy occurred only after Rock had painstakingly evacuated everything from Likiang.

Rock's immediate reaction to the news was severe because he had, in a sense, lost a child. For a few days, according to Paul Weissich, he felt perilously close to suicide. But scholarship, unlike a child, could be brought back to life, and Rock recovered his equilibrium rather quickly. From that time on he directed his passionate attention toward getting back to Likiang to replace the loss. He did not let age, now 60, or ill health interfere with his design; he did not vacillate; every move was a deliberate step back to China.

He did not reach the Orient again until the fall of 1946. This was later than he would have liked but, owing to the reorganization of his several affairs, the delay proved useful. His consultations with the Army Map Service involved only a fraction of his time, and he used the remainder to his advantage. His cash resources were low. He figured he had spent around $18,000 of his own money on Nakhi research since 1930, and Chinese inflation had made big dents in his bank account. He needed, therefore, financial backing for any further field study. Curiously, he did not have much confidence in the power of his unique knowledge to attract institutional support, and he used his library to establish bargaining leverage. He went to Cambridge to discuss his situation with Serge Elisseeff, Director of the Harvard-Yenching Institute, extending again the offer he had made of his books after having discovered the mess in Honolulu. Elisseeff appears to have been more tempted by Rock's scholarship and research proposals than by the literature he had collected, for Harvard-Yenching already had substantial library resources, but Rock insisted on tying everything into one package.[1] It bothered him that most of the books were still in storage in Hawaii, and he wanted someone to pay the freight charges and provide a home for them. Elisseeff, however, had no intention of accepting the collection sight unseen, and because Rock did not have a catalogue and could only describe his holdings in generalities, he refused to commit himself. He made a tentative offer to appoint Rock as a Research Associate, at a salary to be agreed upon, and stated that Rock could bring his library to Cambridge, where storage space would be provided pending an evaluation of the contents.

This left Rock with shipping charges he could ill afford, and he complained about this to E. D. Merrill, his old friend from the Philippines and first trip around the world, who had advanced in life to the post of administrator of all the botanical collections at Harvard, including the Arnold Arboretum. Merrill suprisingly volunteered Arboretum money to pay the costs, a gesture he would have been hard pressed to justify in terms of legitimate Arboretum interests. He was, however, a strong-minded administrator who did not expect to be questioned and, accordingly, he was not. The simple truth was that he liked Rock and admired his work; the man had suffered some bad luck with the loss of his trunks at sea. Merrill commisserated; Japanese bombs had wrecked the library and herbarium he had built up carefully during his years in Manila. Rock asked for help, and Merrill gave it. Merrill also sensed Rock's insecurity in working out the terms of his agreement with Harvard-Yenching, Rock was experienced in dealing with Chinese muletiers, not with university red tape. Merrill advised him of procerdures, and acted as an intermediary. In fact, it was Merrill who warned Rock against relinquishing ownership of his library at that stage because he might have to "eat" it someday; Merrill turned out to be right.

Rock went to Hawaii during the summer of 1944 on an Army No. 2 Priority Flight to arrange for the library to be shipped. Gregg M. Sinclair, the man who had replaced Crawford as president of the University, had long realized his predecessor's blunder and made several appeals to retrieve Rock's library. The tone of Rock's correspondence with him had been friendly, but he had refused to reconsider the situation. Then Sinclair had written Rock in China to ask if he would accept an appointment as Research Associate; the letter was vague, without mention of salary or library. Rock, thinking he would leave China forever, had agreed in principle, but that was before his work for the Army, the loss of his material, his discussions with Elisseeff, and the materialization of better opportunities elsewhere. Though he now declined the position, he accepted the apology implicit in the offer, and disposed of his bitterness toward the University of Hawaii. He organized his library and other possessions for their journey to Cambridge, leaving ample time for the inspection of plants he had introduced to the islands and a pleasant social life. He returned to the mainland.

Rock's obligations to the Army Map Service did not terminate until March, 1945, and the war in the Pacific dragged on until August. Rock haggled Chinese-style with Harvard-Yenching, using Merrill as his go-between and causing the Institute's legal advisor to observe that the man had been out of the real world for so long that he apparently needed some close friend to calm his anxieties.[2] Though Elisseeff considered the technique absurd, Merrill undoubtedly saved the situation for Rock, who could easily have ruined his own chances by blowing up over some trivial misunderstanding. Even with Merrill to soften the blows, nerves on both sides were often on edge, and people at the Institute, like those at the National Geographic Society in days of yore, recognized Rock as a cantankerous customer. The offer to donate the entire library was withdrawn, but the Institute would receive many original Nakhi manuscripts and photostats of others. In the event Rock wanted to sell his books, Harvard-Yenching had first option to buy. Wills and codicils were drawn and redrawn in quadruplicate. The question of Rock's salary was argued for months. He wanted $6,000 per year, traveling expenses, and some assured income after the mandatory retirement age of 65; he settled grudgingly for $4,500, expenses, and a five-year contract from July, 1945, to July, 1950, with no guarantee of anything after he finished his proposed work in China. He worried that Elisseeff was trying to limit the scope of his research in the field and had to be pacified on that account. Owing to Rock's various fears and his circumspect manner of handling things, the negotiations dragged on longer than the war.

Rock also showed Elisseeff a set of page proofs of his history of the Nakhi kingdom and a copy of his 1936 contract with Vetch. After reaching page-proof stage in 1940, the work had progressed no further; Vetch had been drafted, and the Shanghai printers had been left without further instructions. Rock was eager for a finished product, and Harvard-Yenching was sufficiently impressed with what he showed them to agree to finance its publication through the Harvard University Press, a matter, in the end, of $20,800. Lawyers looked at the original agreement with Vetch and declared Rock was not bound by it; poor Vetch later protested in vain.[3] Since the type had to be reset anyway, Rock was requested to make alterations. Rock kept busy, what with editorial work, preparation of maps to accompany the texts, classification of the Nakhi manuscripts that now belonged to the Institute, beginning his *Historical Geography of Northwest China and Northeast Tibet,* which he never finished, and getting ready for China. He behaved more like a man of 32 than of 62, hurrying between Washington, New York, and Maine, the last for some peace and rest after a brief illness; from mapping rooms to banks, from doctors to editors, issuing orders and decisions which occasionally cancelled each other in hasty letters scrawled on various letterheads. Harvard University Press worked quickly—Rock complained without much justification in view of the complexity of setting a text with many Chinese and Tibetan characters—and he had read all the galleys and most of the page proofs for the first volume of the Nakhi kingdom study by the end of August, 1946. He started back to China near the end of the next month. Stopping en route in Honolulu, he was greeted by the usual bevy of reporters. He told them he was returning to Likiang, "where life is not governed by the ticking of the clock but by the movement of celestial bodies."[4]

For a while, however, Rock was obliged to contend with reality in brutal form. His trip to China was marked by a midday skirmish in the Manila Hotel when "gangsters" burst into the lobby and started shooting indiscriminately. Rock was grazed on the ankle while he lay on the floor behind a pillar; the man in front of him died from a bullet wound in the head. Though unpleasant, the incident was out of context; Rock's real troubles started when he reached Hong Kong. His cable to Harvard-Yenching told the story: *Airborne Cost Hongkong About Four Thousand Other Routes Impossible Suggest Your Cabling Ambassador Assist Otherwise Propose Institute Pay Half Myself Remainder.* He had 1,200 pounds of excess baggage: kitchen utensils, household supplies, bedding, linen, medicines, books, kerosene for lamp light, photostatic copies of more than a thousand Nakhi manuscripts, clothing, shoes, paper—all to supply him for three years. He also packed some of the precious gold the Muli king had given him many years ago. "I could sell them in case of necessity," he explained, "so as not to be stranded."

The only alternative route to Yunnanfu, where he had to go anyway, was by truck via Kweichow. This trip, Rock claimed, could take as long as three months because there was practically no gasoline; what there was was frightfully expensive and transported by airplane. If a truck ran out of fuel on the road it might have to wait as long as a week for a new supply.[5] Owing to renewed civil war disturbances, the road was swarming with bandits. Harvard-Yenching had allotted a total of $2,500 for expenses based on an old estimate by Rock and now was totally nonplussed by the sudden necessity for large amounts of cash. Even supposing, as they did erroneously in Cambridge, that Rock's figures were in Mexican dollars, the amount seemed outlandish. Letters to ambassadors would not produce any results, and they already had invested so much in Rock that they could not afford to abandon him now. Reluctantly, they paid.

Since no word arrived immediately from Cambridge, Rock went ahead to Yunnanfu and applied for a landing permit for the plane that would take him to Likiang. The provincial capital was full of war relics and reflections of economic distress. "There has been a black market in Yunnanfu," Rock wrote back to Harvard, "where American Army supplies, left behind after the evacuation, were sold—from uniforms, blankets, to yarn and coffee, but at exhorbitant prices. When I arrived little was left as the foreign community had bought most of it. However, the Chinese did not dump the supplies on the market all at once but, in order to keep the prices up, sold things a few at a time . . . When all was sold out (most of the goods belonged to officials who let them be sold at a commission) the Government posted a proclamation stating that anyone selling American goods on the Black Market would be shot."[6] Chinese currency fluctuated drunkenly. One U.S. dollar sold for 4,300 Chinese dollars when Rock arrived in Yunnanfu; it went up to 7,200 for a while; when he left for Likiang it was at 6,400. Prices behaved in response to currency. Rock, who only had been permitted to take $US 250 from Hong Kong, was naturally afraid to exchange his American currency for the Chinese unless he could spend it immediately. He did, however, change some of his bank drafts for U.S. dollars through American soldiers who were going home. They had been paid in American dollars, "as otherwise they could not live should they be paid at the Government rate; it is expected that they sell their money on the Black Market" which they could not take out of China in cash; but they could leave with bank drafts or travelers' checks.

China's latest version of chaos shocked Rock much less than the sudden appearance of order would have done; he never had known anything but unsettled conditions. With Western self-assurance he clung to the belief that his superior logic could overcome whatever obstacles the disorder might create for him. He also counted, as in the past, on isolating himself from most of the problems in Likiang. His landing permit arrived on the 6th of December and he started bargaining with CAT, which agreed, after much deliberation and double checking, to fly him and his baggage to Likiang for $690 U.S., much less than he had anticipated. He arrived on December 30 in a terrific headwind, and the pilot executed an acrobatic landing on the third try. While pilot and co-pilot stayed behind in the plane, Rock, ashen but relieved, stepped on the airfield at the foot of the Snow Range. That night he spent in a miserable, clammy hut, but the next day he collected his entourage, rented a new house in town, and within a few days was established and ready for work. The house had eleven rooms and was on the banks of a sparkling, spring-fed stream. Rock paid $100 U.S. a year for it, and for an additional small sum, rented three little nearby fields to plant his vegetables. Everyone was very friendly and, as Rock said, his advent in Likiang "was more like a homecoming than arriving in the wilderness."[7]

His old sorcerer had died in August so he hired in his place the departed's older cousin, also a *dtomba*. Life in Likiang had changed very little during the last three years; there was still no electricity, heating, news service, or running water; the mail arrived twice a month. The people's problems had not been solved, and some new ones had appeared. The cattle had been hit by an epidemic, and humans now hitched themselves to plows like oxen to turn the earth. The missionaries who had evacuated when the Japanese made their last desperate push in Kwangsi in 1944 had not returned, owing to the civil war and transportation difficulties. The only foreigner besides Rock in the village was Peter Goullart, "and he is half crazy". "I might as well be living on the moon," Rock said.

Unlike the moon, Likiang was not far enough removed from the mainstream of Chinese activity to have escaped the financial crisis. Though prices were lower than in Yunnanfu, money immediately became a major problem for Rock. He had left nearly all of his life's savings, about $35,000, in a Yunnanfu bank when he went to Washington in 1944 because he had not

been allowed to take the money out. He had assumed, however, that the earlier inflation was a wartime phenomenon that would end with the Japanese occupation and that, wherever he was, he would be able to withdraw the sum. The expected financial normalization had never occurred, and Rock worried over the fate of his deposit and his future. "How long I can stay here, unless conditions improve, I don't know," he wrote Merrill in January. "I can assure you that it is indeed the love of work that is keeping me here, especially as I can't get medical relief . . . When I left the States, there was hope of peace and stability, but this is not materializing."[8] By mid-April, 1947, it was costing him $12,000 Chinese simply to post a letter to the United States.

By summer things were even worse. "Life, on account of the economic situation, will soon be impossible here," he advised Elisseeff, "with the present exchange for the U. S. dollar which . . . is $11,640 Chinese; the transfer of money from Kunming [Yunnanfu] is very expensive, 10% on account of the bandits on the road, telegrams are $10,000 a word from Kunming to here, so by the time one gets money sent here there is little left of a million dollars, and that amount is all that can be sent at one time. A million lasts about ten days. . . . The other day gold was $3,600,000 the oz., today it is five million. You can see what the U.S. dollar should be worth, but isn't."[9] The American dollar stayed at a fixed rate while the Chinese dollar purchased less and less. "Beggars curse you if you give them $50 [Chinese]," Rock remarked to Merrill.

There was nothing Rock could do about the Chinese economy except worry and solicit the sympathy of friends who lived in lands of predictable currency. Meanwhile, he had work to do. His new *dtomba* proved, on experience, to be a faker who knew less than his employer about the ceremonies he chanted. It took Rock three weeks to discover his weaknesses and relieve him of his duties. His replacement, a young sorcerer, provoked Rock's remark to Merrill that the "younger generation knows nothing,"[10] before he, too, was discharged. On the third try Rock finally found a *dtomba* who satisfied his needs; in fact, the best he had ever had. The man was intelligent and willing, and Rock enjoyed working with him even when he repeated work done years ago.

But he had gotten off to a slow start and there were many demands upon his time. "There is, of course, no hospital, no clinic, no doctor," Rock explained to Merrill. "People die, and there is no reason for them to die, for they could be helped. Nearly all afflicted with acute appendicitis die, for there is nobody to operate and the first thing they do is take a strong laxative, which kills them. . . . They come to me with all kinds of ailments, and I try to help them the best I can." Rock did not have the heart to turn the sick away when they ventured into his courtyard, but the expense, both real and in time and energy, was high. When Goullart had to travel to Yunnanfu in April, Rock charged him to bring back a large supply of medicine. Penicillin had become outrageously costly, and Rock could not afford any more and had to go without. At the end of May he became weary and discouraged by his patients and took a week off to camp in the meadow below the Snow Range and refresh his energies.

Not wishing to be considered derelict in his research, Rock did not mention his medical efforts to his Harvard sponsors nor did he tell them of his own health problems, which, by June, had taken a new turn. His insides, which had always been the source of his troubles, behaved themselves for a while, but he was now afflicted with excruciating shooting pains on the right side of his face. There were days when he could not eat solid food. "Every bite," he confided to Merrill, "is like a daggar's thrust from the palate to the top of the head." He consulted his medical books and emerged with a diagnosis of "causalgia"—facial neuralgia —which he learned might be treated by an injection of procaine to create a nerve block. Since he had procaine and a hypodermic on hand, he wired a doctor in Yunnanfu to inquire about the proper administration of the drug and was informed that only an expert could do the job. A trip to Yunnanfu by air would cost too much and by motor would be too dangerous because of bandits; he was trapped again. The pain kept him awake nights, interfered with his work, and drove him nearly crazy. He halfheartedly wished he had stayed in the States or even in Yunnanfu, where he might get medical attention. At the same time, his fear of being ordered out of China and dropped by Harvard-Yenching because of ill health was strong enough to prevent him

from saying a word about his condition to Elisseeff, and he persisted mulishly with his work. No physical ailment had ever distracted him so totally. The neuralgia came in sieges of several days and then vanished without reason; when the pain was present he could not even wash his face, much less shave; when it was absent, he dreaded its return. During the summer his stomach acted up and, since he had hardly been able to eat, his weight dropped sharply. In November he developed an inner ear infection, which caused nausea and vertigo. Rock doctored himself as best he could, remaining in bed only as a last resort. In the old days he had sometimes relished the role of invalid; now, at an advanced age, pride and energy cooperated to keep him going in a more or less normal routine.

The Musée Guimet in Paris agreed to publish his book on the Naga Cult of the Nakhi tribe, and he finished off a manuscript for them in the fall.[11] The Catholic University in Peking was interested in publishing some of the other religious ceremonies, and he had a nearly completed manuscript for them, too. With the help of the excellent *dtomba*, Rock made scholastic advances, compensating for his drowned notes, yet he could not sustain a sense of accomplishment; it was too dependent on his physical condition. But he always had been moody and easily depressed, often for lesser reasons, so his psychological condition was, for him, normal.

INFLATION AND ECONOMIC DISTRESS

It was also characteristic of Rock to be able to distinguish subjective misery from external events and to retain his power of observation even in his bleakeast moods. Very little escaped his scrutiny. His most immediate concern after his work was China's infinitely collapsable currency which showed no signs of levelling. By the end of June, 1947, he often found himself spending $100,000 a day for eggs, charcoal, and meat. "Bad to worse," his expression for the Chinese economy might have been stated with justice as "worse to awful." Ten thousand dollar notes had appeared in Likiang, causing prices to rise immediately by 50% and retiring the abundant supply of $10 and $20 notes into obsolescence. The Likiang bank, stuck with the now worthless denominations, unloaded them upon the public by forcing people to accept one-tenth of any sum withdrawn in $10 and $20 notes and refusing to accept them for deposit. Prices doubled sixty-seven times between January, 1946, and August 1948.[12] Rock, sinking deeper into a financial mire every day, preferred to sell his Muli gold, which came closer to reflecting the real value of the Chinese dollar, to his American currency which was still exchanged at a ridiculous fixed rate. But his supply of gold was exhaustible and his financial optimism exhausted. "I shall be absolutely broke when I get out of here," he predicted, "and that at retirement age."

Economic desperation unleashed stealing and corruption on a scale Rock never before had witnessed in Likiang. Soldiers stole vegetables from his garden at night and, along with the squirrels and insects, left little for his table. "There is no redress," he wrote. "The local authorities are afraid of the soldiers. They go into people's houses, take the food out of the pots, take the bedding from the womenfolk, and simply walk off. Meat is not sold on the open market (there are no meat shops) but in hidden spots for fear of the soldiers who openly would grab what they can and walk off. These are the 'protectors' of the people who pay taxes to support them." Highway robbery had become an established fact of life, and it was forbidden to send money through the mails so as not to endanger the mail carriers. The Likiang postmaster, however, used Chinese bank notes from the postal savings bank to buy silver dollars and regularly sent the mail carrier to Muli with the silver to trade it for gold. As no one interfered with the postmaster's activities, it was safe to assume that higher authorities shared in his profits.

The peasants' attitude to official abuses riled Rock; it seemed to be the same indifference he had always known. " . . . All they say is 'mei-yu fa tzu' = there is nothing to be done. That's why they have such a rotten government. It's the crooks that have the initiative. . . . There is not the slightest public or community spirit; all for the family. If a neighbor is robbed, so much the better; it was not they who were robbed. The highest aim in China is for a man to become an official so he can squeeze the people. . . . I have never in my life seen as much corruption as there is in this country. Laws are on paper only and are only for those who haven't the go in them of a worm—[which], at least, squirms when stepped upon. All the officials smoke opium and gamble. It is they who support the black market . . . the higher the official the bigger the crook. The only reason why foreigners are still head of the customs is because if the Chinese were in charge, Chiang Kai-shek wouldn't have any revenue.

"I hope the British will *never give up* Hong Kong. It is the only decent spot in Asia now. . . . Chinese in Hong Kong have written me 'we are glad we are not living in China.' To call China one of the Big Five is worse than nonsense. They live on their reputation of 2,000 years ago. They have no *present* culture. What culture they had was centralized in the capital. The rest of China was, and *still is* a grand pig sty where slavery and corruption, banditry, inefficiency, and deceit [are] paramount. All is face, like the houses of the wealthy—a lot of carvings and gilt, yet no water, no toilet. They go on the hillside. You see them sitting in a row every morning . . . Their morality, because they marry early, is a myth. When they are sixty they still buy girls to be their concubines. Women are bought and sold like pigs; they have nothing to say."

This was Rock's old diatribe, the complaints he had registered for twenty-five years made more bitter by the monotony of repetition and the apparent failure of the browbeaten peasants to do anything but cringe. Rock had always believed that human beings could be pushed only so far before they would rebel, but now he doubted. Nakhi peasants in Likiang—as poor people throughout China—had been abused beyond the limits he recognized as tolerable, through civil war, communist marchers, provincial infighting, and Japanese aggression; by bandits, opium dealers, magistrates, tax collectors, and soldiers representing various authorities. They had starved in famines, drowned in floods; had been sold into slavery, received beatings from soldiers, and had been squeezed by all officialdom. Yet they submitted themselves without protest, almost inviting abuse by their very meekness. It seemed to Rock, after a quarter of a century in China, that the country could never change because its people had accepted subhuman conditions. He misjudged the situation, measuring their submissiveness on a Western scale, and assumed that if the peasants had not rebelled by now, they never would. Therefore he could not take the communist offensive seriously. Only a few Westerners who knew China, among them, his old friend Edgar Snow, would have disagreed with his evaluation.

Communists and Nationalists vied for power in northeast China throughout 1947. The scene of their battle was so remote from Yunnan that, for the people of Likiang, the civil war might have been fought on the other side of the world. They were struggling to survive the indignities of inflation and squeeze and, being politically naive, did not divine connections between $10,000 banknotes and military maneuvers in Manchuria. But, as Fairbank pointed out, the financial crisis eroded public confidence in political authority. "Life under hyperinflation is a slow strangulation," he wrote. "Salaries and wages never keep up. Furniture, books, and clothing go for food. Gradual malnutrition produces skin diseases, stomach ailments, tuberculosis. The whole society sickens, and the responsibility is put on those in power." Every time prices rose, public faith declined. A whole nation was groomed for political change—not necessarily a communist government but *any* power which could get them off the economic treadmill, something the Nationalists did not seem either willing or able to do.

The obvious rhetorical question, after the events could be examined by historians, was why Chiang Kai-shek did not launch a full scale attack on the economy instead of on the reds. The Generalissimo clearly misjudged his priorities, as he had done in the 1930s by choosing to fight the communists before the Japanese. While it is very simple to see his errors now, the situation within the Chinese context was not at all transparent. Chiang acted in the honorable tradition of the Chinese warlord, seeking first to eliminate rivals and consolidate his power. He knew the need for reforms but put them off. He was, in a sense, a captive of his culture. Immanuel Hsu believes that "the Nationalist officials continued to live under the shadow of the Confucian distinction between the rulers and the ruled, and looked down upon the peasants as an inert nonentity."[13] That, in the 20th century, eighty percent or more of the population could be lightly dismissed as no more relevant to political power than water buffaloes seems to us ludicrous but, with the stunning exception of the T'aip'ing rebellion in the mid-19th century, the Chinese peasants had been more inert than not. Surely Rock did not count them as a political force; the Americans who sent aid to Chiang Kai-shek did not take them seriously; Stalin and his clique in Moscow underestimated their revolutionary potential; Chinese communists in earlier years had argued bitterly among themselves whether the peasants could wage a successful rebellion. Only Mao recognized the agrarian masses for what they were: China's greatest natural resource waiting to be tapped. By drawing on that power, Mao played with a whole new set of rules, totally alien to Chiang's conventional thinking. Meanwhile, Chiang inadvertantly played right into communist hands.

Theoretically civil war might have been averted. In April, 1945, before World War II had ended but when the Japanese had already shifted into reverse, the communists proposed a postwar coalition government to the Kuomintang. Negotiations on this subject were held intermittently for over a year, sometimes with American mediators, first Ambassador Patrick Hurley, later Secretary of State George Marshall, against a backdrop of escalating warfare in the northeastern provinces. In their determination to have a strong, unified, democratic and, especially, friendly China, the Americans believed in the possibility of a political settlement much longer than either Chinese faction; Rock's original optimism for a peaceful, stable China had been generated by American newspapers. But Chiang had no intention of clasping the communist viper to his bosom. He mistrusted the reds deeply, fearing that they would be more dangerous to his power as partners in government than they had been as enemies in the field. He can hardly be faulted on that account; he and his ruling clique were not among the Chinese communist ideological objectives. More essential to Chiang's rejection of a coalition was his hatred of an adversary he had fought for almost twenty years and which always had contrived to elude his more numerous and better supplied forces. He could never have accepted more than token communist representation within his government; any solution that reflected their real power—and they would settle for nothing less—was unpalatable to him. As for the communists, they regarded the possibility of a negotiated settlement as highly unlikely but they were not insensitive to the good impression which their offer made internationally and among China's many disenchanted, war weary, inflation fatigued millions.

There is one more important consideration: Chiang thought he could beat the communists on the battlefield. Of course, since he had often believed this before and had been proven wrong, one might imagine that he had misgivings, but he did not. On paper, at least, he had all the advantages of a larger army and a superior arsenal. The precise measure of the Nationalist advantage was, however, also inflated; Kuomintang forces were actually fewer, and communist greater, than reported. This overestimation perhaps originated in ancient Chinese etiquette: it was considered polite to tell people what they wanted to hear, and in this case the Generalissimo certainly wanted to hear that communist forces were weak. Either he chose to believe a courteous delivery of information or there was some inaccurate counting on the Nationalist side, but even with the distortion of numbers, the Nationalists still came out ahead. What could not be calculated was the spirit of the troops, and Chiang's forces were demoralized and disorganized.

Rock observed the low army morale even in Likiang in December, 1947. "The Chinese soldiers of Wei-hsi," he informed Elisseeff at the time, "have mutinied and left with their rifles and even machine guns and have come to the neighborhood of Likiang. They have even attacked a group of soldiers 40 miles from here, killed two, and disarmed the rest. Two are severely wounded and are in a house or barrack here. The soldiers, who get $6,000 Chinese dollars a month, have not been paid for six months. They get one meal a day and cabbage soup at night, have no bedding (they sleep in straw) and not even straw sandals. They go to the farms and tear off the fibrous bast of a palm that grows here and make shoes. . . . Their officers use the salaries of the soldiers to do business. They are forced to go to vegetable gardens and pull up what they need."[14] What happened in Likiang was repeated, with variations, by Kuomintang soldiers throughout China. Needless to say, the civilian population did not find these soldierly exhibitions exhilarating. As the fighting with the Reds wore on, Nationalist troops defected or deserted and seemed bewildered by the communists' guerilla tactics.

In January, 1948, Chiang Kai-shek publicly expressed confidence that he could eliminate most of the "bandits" within a year—he still disdained to identify his enemies by their official title—even though the Peoples Liberation Army had moved to the offensive, and there was dissention at high levels in the Kuomintang. Mao, meanwhile, exuded optimism and scorned the Nationalists' greater numbers and the massive aid they were receiving from the United States. American efforts at mediation had failed, and both sides had committed themselves to annihilating the other. Only one man could be correct in his prediction; either Chiang or Mao would have to choke on his words.

Rock had no thoughts for politics at the beginning of 1948. His physical afflictions were much more severe than ever before, and he became totally absorbed in pain. His digestion was bad and the facial neuralgia constant. He stayed on a liquid diet during most of January and February; as a result he lost a great deal of weight but did not cure himself. One of his correspondents in Boston, a doctor, wrote him, "If you want my professional advice, I say tersely 'Get the hell out of there.' "[15] By the third week of February he realized he could not continue this way of life and expect to live very long. He wired Chennault via the Civil Air Transport in Yunnanfu and asked that a plane be sent for him. It arrived on the 29th, collected Rock, one assistant and a suitcase, and made Yunnanfu in an hour with the benefit of a good tail wind. He had left all his belongings in Likiang, with every intention of returning. "Don't think me crazy," he warned Merrill, "unless the Reds get here."

Rock received some medical advice in Hong Kong which relieved his intestinal distress. He had been treating himself with daily enemas in Likiang and was told to give them up and take Epsom salts instead, but the neuralgia involved more drastic measures, so he proceeded, via Europe, to the United States. Though he was in pain nearly all the time, he permitted himself more than a month in Europe to discuss his publications with the Musée Guimet in Paris, to give some lectures, and to accept the Stanislaus Julien Award of the Institut des Belles Lettres of the Académie Française, as well as another honor in Switzerland. He did not go to Vienna but got in touch with Robert for the first time since Hong Kong and asked if he would like to come to Likiang and be his secretary; he felt deeply guilty over Robert.

The nephew answered affirmatively, as though nothing had happened, and his uncle promised to arrange everything.

Rock flew to New York in May and, after various medical consultations, found himself in Massachusetts General Hospital at the end of the month. The nerve block he had hoped would solve his neuralgia was only a temporary treatment which would have prevented his return to China; cutting all the ailing nerves would have involved a long period of observation. The doctor advised him, therefore, to accept a compromise operation, but he did not understand until after it was over that the neuralgia could recur in the area of his eye. The doctors performed their work with local anesthesia only, and Rock was conscious for the whole thing, terrified every minute. When it was over, his lips and palate were numb. The nightmare of the surgery, the deadness of half his face, and the possibility that he was not even cured, drove him to a frenzy. "I nearly jumped out the window," he told Merrill, "but I want to die in China, not here." The Western world held nothing for him anymore. He felt himself "quite apart from the common herd . . ." even in Europe; as soon as he recovered, he hurried back to China, leaving his doctors, sponsors, and friends hovering between admiration for his courage and doubts of his sanity.

During the four months or so of Rock's absence, the communists gained ground in their contest with the Kuomintang, and areas beyond the battlefield were beginning to reflect the struggle. Communist organizers, some of them left behind during the Long March, were active in key locations, teaching the ways of Mao and Marx and Lenin, preparing peasants and students to oppose the Nationalists, soliciting the defection of disillusioned government soldiers. Staggering under the pressures of inflation, the peasants saw some hope; students responded with idealism and energy; soldiers crossed over. When government troops in Yunnanfu arrested four communists, the students of Yunnan University demonstrated in the streets to demand their release and forced a confrontation with the soldiers. Minor disturbances were in progress for several days after Rock arrived on his way to Likiang, and Nationalist frustration finally boiled over. The day Rock's charter plane was due to leave for the west, soldiers surrounded the university dormitories at 4. a.m. and opened fire with rifles and machine guns. The students, evidently alerted to the attack, fled to the rooftops and threw bricks, rocks, and nitric acid on the soldiers; a few students were armed with revolvers. "Girl students threw soldiers out of the windows," Rock heard with surprise, for girls in China had always been meek and obedient. His flight left at dawn, and he did not wait for a casualty count. He felt discouraged and did not really care who had "won" the encounter.

He was also bothered by the visible resurgence during his absence of anti-foreign animosity, specifically directed against Americans. Since many students sympathized with the communists, and the United States had thrown its weight behind Chiang Kai-shek, the anti-American sentiment followed logically; not so, however, the argument. Rock tried talking with students in Yunnanfu. "They say we are helping to arm the Japs and put them economically on their feet to the detriment of China," he wrote Merrill. "I told them that we taxpayers cannot be expected to feed the Japs forever, but they must be put on their feet again so that they can sell their produce and buy food." His explanations in accented Chinese made no headway; the students would soon be using a new argument anyway. The American Consul in Yunnanfu warned him that all Americans might have to evacuate Yunnan in six months. Rock refused to believe him but was uneasy about the future.

Likiang, one step farther removed from the center of the political whirlpool, was quieter than Yunnanfu but also troubled. Nationalist authorities suspected the presence of communist agents, though nobody knew who they were, and incidents formerly blamed on bandits were now officially reassigned to the reds, while apolitical robbers still flourished. Ten boxes of rifle ammunition disappeared one night from the old temple which served as military headquarters. The soldiers caught six of the thieves and retrieved three of the boxes. No one could decide whether the offenders were communists or not. It hardly mattered. The colonel in charge in Likiang, who had committed some offence that had been reported to Yunnanfu, and who was about to be replaced, smuggled four more boxes of ammunition out of the barracks and sold them for $400 silver a box. Disorder and desperation had reached the level where everyone acted in the interest of self-preservation. Paper currency stood at six million to one U.S. dollar and still rising, and no one would accept paper money in exchange for goods. People in a position to do so hoarded silver, which became scarcer every day. The Kuomintang soldiers' lot had grown worse. A sergeant earned one million a month plus his rice, but no vegetables or meat. A cattie of pork cost $265,000. Sometimes the soldiers were not paid for months. "The situation is hopeless," Rock reported, "and something is bound to happen soon. . . . Can you see any hope for such a country?" It was, obviously, not a question to which he expected anything but a negative answer.

Rock could only imagine one solution and that on a provincial level. "I presume if it comes to the worst, Yunnan will declare its independence from the rest of China," he told Merrill. He still had not grasped either the strength of the communist thrust or the loss of Yunnan's idyllic isolation. He made a natural mistake based on historical evidence, but old formulas no longer applied. Besides, the provincial government was in shables of its own. Chiang had appointed one Gen. Lu Han as governor in October, 1946. Lu specialized in stealing what he could of the ex-governor's silver and gold and in distributing provincial posts among his corrupt cronies. Lung Yun, however greedy, had kept relative order—he had been demoted to head of Military Headquarters in Yunnanfu; Lu Han paid no attention to anything outside the immediate vicinity of the capital.

By mid-August, when the exchange rate went up to 12 million Chinese dollars to $1 US, the Government finally tried some currency reform. All gold, silver, and foreign currency were supposed to be turned over to the government, and a new "gold yuan" paper currency was issued at the rate of four to every U.S. dollar. People, however, were understandably wary of paper money; in Likiang they did not give up their silver. On September 27, 1948, about a month after the new system had been initiated, Rock reported to Egbert Walker: "No amount of proclamations have been posted here which very few people can read as there are only a handful of Chinese here, and most of the Nakhi can neither speak nor write Chinese, so the walls are very patient and so is the paper. The people have shut up their shops rather than accept that money in payment."[16] Six weeks later he supplied a further analysis to Merrill: "The new currency is causing chaos; it is going downhill fast . . . Unless you have local silver, you starve. . . . Even the Government tax office refuses to accept the new notes and demands silver which the Government has declared illegal." People in Shanghai had been shot for handling silver; in Likiang the Government would not accept anything else.

It was at this point that the peasants of Likiang reached the limits of endurance and started to shed their customary passivity. They turned first against the commanding military officer. The problem of the soldiers had become impossible. The troops were paid in the worthless currency which both the peasants and the merchants refused in exchange for food; the soldiers, therefore, stole—directly from the peasants rather than the merchants—in order not to starve. One day in late October a mob of farmers, armed with everything from spears to machine guns, appeared outside the military headquarters and threatened to kill the commanding officer. The latter ordered his troops to open fire into the crowd with machine guns, but they turned their backs on the crowd and pointed the weapons directly into his quarters. He decamped for Yunnanfu with impressive haste. Rock witnessed the near riot, remarked that the situation was "fraught with danger," and continued to wonder about rumors of "an influx of reds." In his sympathy for he peasants' and soldiers' grievances, he overlooked the evidence which indicated calculated communist organization behind the demonstration. Peasants armed with machine guns, rifles, and aggression did not spring out of the ground overnight like mushrooms after a heavy rain. In another six months Peter Goullart would be surprised to discover dedicated communists among men with whom he had worked in the cooperatives. Neither Westerner, however, recognized the early signals of red involvment in Likiang.

One by one officials in the district collapsed or vanished. "The chief justice and the prosecutor of the local high court made off with all the silver they squeezed," Rock wrote Walker. "They left cases undecided and skipt [sic] with the loot. On the district border 30 students held them up and they had to cough up some of the loot before they could proceed." Judges and military officer gone, tax collectors and postal official corrupt, effectively, administrative authority in Likiang was near the breaking point. Similar conditions were reported in other towns throughout the province. Bandit hordes swelled, roamed the countryside, and plundered at will. The robbers, many of them disaffected soldiers and peasants, were as desperate as anyone else. Communist guerilla groups were also thought to be operating in the province. "The issuing of this new money just accelerated the collapse which is bound to come," Rock predicted.

The commotion hardly encouraged scholarly research. By the end of the year Rock admitted against his will that his days in Likiang were numbered, and he scrambled to make the best use of whatever time remained to him. He abbreviated notes and demanded long hours from his sorcerer and secretary. He sent for Robert, thinking that another assistant would hasten his progress and ignoring the potential danger to his nephew's safety. The tractable Robert agreed to come but, fortunately for him, was stopped in Paris by visa difficulties and had to return to Vienna. Never in Rock's experience had China been in such an uproar and never before had Likiang been really disturbed. Yet now, of all times, Rock was reluctant to leave. His financial resources dwindled rapidly owing to inflation and the bad exchange rates; he had, furthermore, borne the entire expense of his trip to the West in 1948. He sold gold and silver relics piece by piece. He convinced himself of an obligation to Harvard-Yenching to complete his promised work, but no one in Cambridge would have blamed him for leaving China. Elisseeff and Donham, in fact, would have been quite relieved to see him out. What appeared as a martyrdom in the name of scholastic honor was, however, a disguise for more primitive emotions: Rock had nowhere to go. He thought of China as home now and could not imagine a life anywhere else.

THE COMMUNIST INVASION

Chiang Kai-shek chose his New Year address of 1949 to dignify the enemy by calling them communists instead of bandits. This delicate verbal transition originated in the Kuomintang's indelicate military position which was, approximately, hopeless. In an effort to draw strength from his Western allies,

Chiang displayed a sudden fondness for a mediated settlement; the communists, confident of their strategic superiority, issued strong terms which reflected their advantage and effectively eliminated the possibility of a political solution. Nationalist defeat filled the air, and Chiang resigned as President, in theory if not in practice, on January 21. By the end of the month the Peoples Liberation Army held almost all of north China, including Peking. Civil war continued through 1949, but it was played out like the endgame in a chess match in which one side had the advantage in pieces and position and the other side, praying for a stupendous blunder, refused to resign until the irrevocable checkmate. The blunder never came.

The Nakhi peasants in Likiang did not stage any further demonstrations but adjourned to their farms; the village itself was quiet as concentration shifted to the robber bands which constituted the most immediate threat. "This province, with the exception of Kunming and a few other places, is practically in the hands of bandits,"[17] Rock advised Elisseeff in early April. "The suburbs of Kunming are not safe, and the post office has discontinued delivering mail . . . Last night there was a scare in this wall-less town, for rumor had it that the robbers were . . . to the west of us, and people living in the southern suburb of this town packed up during the night and left for safer quarters. Hoching, 30 miles south of us, was looted by Chinese soldiers stationed there; the magistrate here had a wire saying they were on their way to Likiang . . ." The rumors in Likiang were wild; people in the streets discussed 100,000 bandits on the rampage. Though he recognized inflated numbers when he heard them—the actual count was more like 5,000—Rock became nervous. He kept in close touch with CAT; Chennault promised him that a plane would be dispatched whenever Rock requested one.

This bandit outfit had real imagination and posed as communists. Having sacked Hoching and Yungpei, they approached Likiang and sent a message to the magistrate demanding that the citizens relinquish their arms, furnish money, and go over to the reds. Unlike Rock, the Nakhi were not bamboozled by this bluff; they sent spies to evaluate the wreckage and Yungpei and reached the inescapable conclusion that the bandits were bandits and not reformers or revolutionaries. Likiang decided to fight the pending invasion, and all able-bodied men and women volunteered their services. The magistrate ordered trenches dug and called for support from the district, even Lolos and Tibetan bandits from Chungtien. Rock stayed awake all night on April 28, waiting for the fight to begin, but all remained quiet. Tension mounted for the next few days, and the gossip, baseless or not, was hardly reassuring. Rock succumbed to his anxieties after a week and wired Chennault to send the plane. He flew to Yunnanfu in May with most of his belongings. Ironically, he had underestimated his Nakhi hosts; they held the bandits off from Likiang and chased them from the district.

During the month of May the Peoples Liberation Army took Wuhan, May 17; Sian, May 20; Nanchang, May 23; Shanghai, May 27. The Government's defeat had become only a matter of time, but Rock still would not believe that all of China would fall before the red juggernaut. The absolute control threatened by the communists was not a Chinese experience as he understood it. He could not, would not, face the necessity for leaving China but hung on resolutely in Yunnanfu. He learned of the deliverance of Likiang from bandit hands when the head of the robber chief was brought to the provincial capital for display. Caravans from western Yunnan reported that Likiang was peaceful, whereas Yunnanfu seemed like a good target for the communists. Rock took the first opportunity to return to his battered Nakhi dreamland. On June 22 he made an initial try, but clouds covered all the mountains in the area, and the pilot could not make a landing and had to return with his disappointed passenger. A second effort, on July 2, succeeded. Rock stepped onto the airfield to Goullart's wry greeting: "Welcome to the Red paradise!" He nearly collapsed from shock. He faltered, looking back at the aircraft, and then ordered his Nakhi servants, who were also on hand to meet him, to unload his supplies.

As they proceeded from the landing field to a house where they would spend the night en route to Likiang, Goullart listed the events that had occurred in Rock's absence: The establishment of a Communist Executive Committee in anticipation of the Red Army, which was approaching from Szechwan and Kweichow; the arrest of the local magistrate and a number of village elders; the "liberation" march, during which a downpour had reduced the improvised images of Stalin and Mao to a soggy mess. Knowing Rock's dislike for the communists and sharing his antipathy, Goullart had tried to forewarn him but had not been permitted access to the telegraph.[18]

From within the half-way house, the two Westerners listened to the roar of the engines as the aircraft took off on its return voyage. Rock reported to Merrill: "The plane which brought me had hardly taken off again when 30 armed men appeared in the little house. . . . They came with rifles and bayonets and occupied all the doors of the house and courtyard; they planted a machine gun in the main gate pointing into the court of the house and then started searching my belongings; they were looking for arms. I had none and so they left as they had come, without taking anything. My cook got cold feet and left in a hurry, a man I had trusted and done a lot for. The others remained faithful. Next day I secured carriers in the same village to take my supplies to Likiang. They were at first afraid and would not move, only after my writer persuaded them, they agreed, and so after having to pay them about ten times what I would have had to pay in ordinary times, we arrived unmolested in my house where I found everything in order."

Rock had observed the changing fates of warlords and political factions; the bandits who became generals, the generals who turned bandits, the assassination of feudal princes, the pitiful attempts of the Nationalists to exercise authority. What he had not seen was any evidence of stability, and this alone gave him hope that the current political manifestation would follow the course of others and disappear like a bad dream. He wanted desperately to stay in Likiang. The house and town were the closest approximations of home for him; his servants had become his family. Gingerly testing the new order, he summoned the Nakhi sorcerer who had been helping him with translations, but all the sorcerers, including Rock's assistant, had gone underground because the new regime had already started smashing temples and sacking lamaseries, enforcing Lenin's dictum that religion is the opium of the people.

A week in communist Likiang made Rock jittery, and his chances for peaceful residence there looked progressively grimmer. "It has been a reign of terror," he wrote a friend in the United States. The peasants "opened the jails and freed all prisoners. The jail is now filled with the well-to-do, and the executioner is having a busy time. It reminds me of the days of Paris . . . during the reign of the Jacobins, as related in history and by the guides who take you through the Conciergerie where Marie Antoinette was imprisoned . . .

"So far I have been left alone, but as I said there is great tension. They are using trenches dug during the last war at the airfield watching for possible planes. So far the mail and telegraphs are functioning . . . They are singing and dancing all day and executions are being held while the people sing to the tune of 'Ach du lieber Augustine, Augustine! . . .

"I am wondering what the next weeks will bring. Proclamations have been posted saying the U.S.A. is the great enemy of the Chinese people, but the lives and property of the foreigners will be respected. I only hope it will hold good. It is very true the poor peasants have suffered a great deal at the hands of the officials, and even a worm will turn when stepped upon. . . . nobody could have endured what they have endured. Now, of course, they have the upper hand and they are after those who have oppressed them."

Goullart, as a representative of the Chinese Industrial Co-operatives (INDUSCO), had worked closely with the local people and had friends in Likiang.[19] As a foreigner and employee of the Nationalists, he was now attacked by the new officialdom as an imperialist agent. The communists confiscated the machinery, tools, and funds he had managed to scrape together for his modest cooperatives and boasted that they would create their own bigger and better cooperatives. Called before the omnipotent Executive Committee, Goullart was surprised to see among its membership some Nakhi men he had never seen before and surmised that they had been trained outside Likiang as advance men for the People's Liberation Army. Such evidence of forethought impressed and discouraged him. Stripped of funds and equipment, and lacking support from the beleaguered Nationalist regime which had sent him to Likiang, Goullart gave up his work. Like Rock he had become deeply attached to the town, but each day the foreigners' situation there became more perilous.

During the first three weeks of his stay, Rock dared venture into the streets only twice. Both times the local population greeted him with sneers and catcalls. Once almost a potentate in this town, he had become a despised Westerner. "They have borrowed foreign clothing from somebody and impersonated Roosevelt in a burlesque, anti-foreign to say *the least*," he reported to Merrill, July 24, 1949. Rock's servants were called slaves of the foreign devil and had legitimate reasons to be afraid. By the third week of July, Rock's nervous system began to falter and, with Goullart, he made the decision to leave.

Arranging for a plane proved more troublesome than Rock had supposed. Writing in German, which he knew nobody in Likiang could read, he managed to get word to the American Consul in Yunnanfu who passed on the request to Chennault and the Civil Air Transport. The greater difficulty come in securing the permission of the Likiang authorities to land the craft, and this they agreed to only on the condition that it would bring the money that the provincial government owed the Likiang school-teachers.

Fearful of last minute detention, Goullart publicized his trip as a temporary absence to pick up a new consignment of medicine. His old friends, however, saw through his bluff, as did the communist officials who had more reason to be thankful for the foreigners' departure than to prevent it. Faithful to his masquerade, Goullart packed only essential clothing, books, and a typewriter, leaving behind many accumulated belongings. Rock, by contrast, read the situation perfectly and packed everything of value, including a substantial library, photographs, treasured artifacts, and clothing. In the end he sacrificed only his furniture, crockery, and a year's food supply which would have been a nuisance to carry.

The day before the plane was due it poured rain, but Rock and Goullart made their individual ways over flooded roads 45 *li* (15 miles) to the village by the airfield. Rock, accustomed to starting early, arrived at the little hut first. Both men were soaking wet and spent the evening trying to dry their clothing by the fire while armed villagers guarded the house. The following day, August 3, 1949, was perfect flying weather, and the refugees stationed themselves at the edge of the airfield at daybreak. The hours were long, and patience wore thin as they listened for the roar of the engines and searched the horizon for their carrier. When nothing rewarded their vigilance by sundown, they went back to the village, afraid that something had gone wrong. They had been inside the hut only a few minutes when they heard the plane land; they dashed back to the field. The chests of silver for the schoolteachers were unloaded, and Rock and Goullart boarded quickly. Just before dark the silver Dakota taxied down the meadow, lifted off the ground, tucked in its wheels, and began to climb over the mountains. In a few hours they were in Yunnan-fu.

Once he determined to leave Likiang, Rock had no intention of remaining in China—at least not for the time being. Americans had been warned to evacuate Shanghai; a town sixty miles from Yunnanfu was "liberated" on August 5, and the communists advanced every day. Only the very few foreigners who enjoyed the favor of top communist officials could hope to live on safely in China; Rock was not one of them. On August 12 he flew from Yunnanfu to Hong Kong. The Yunnan capital fell to the communists about two weeks later.

"I will see how things go during the next year," he wrote Merrill, "and if all is O.K. will go back to Likiang to finish my work, . . . I want to die among those beautiful mountains rather than in a bleak hospital bed all alone."[20]

EPILOGUE

Rock could not believe that China had closed its doors upon him permanently. He assumed that Chinese disorder and apathy would eventually reassert themselves and he would be able to go back to Likiang to work and, one day, to die beneath the almond and peach trees at the foot of the Jade Dragon Mountain. After the communists chased him from Yunnan, he went to Europe and discussed the prospects for continuing his work with Giuseppe Tucci, a noted Tibetanologist and president of the Italian Institute for the Middle and Far East. Then he repaired to Kalimpong, India, in the Himalayas, poised in every fiber for the moment he could return over the mountains to the only place where he wanted to live. He kept his vigil, with occasional interruptions for visits to Europe, for two years, but every day his prospects diminished.

It was, finally, the entrance of the People's Liberation Army into Tibet in October, 1950, and the subsequent establishment of Chinese Communist authority there that forced Rock to accept the absoluteness of his exile. If Tibet, the impenetrable, remote, and mysterious, could be subdued by the new government, then there were no more sanctuaries for him in Likiang, Yungning, or Muli. Chinese participation in the Korean War certainly did not encourage him either. By the summer of 1951, he abandoned his watch, his heart empty and anxious. "I think a lot has passed never to come again," he wrote Andrew Tse, a friend; "at least not in our generation. I hate this place now. The rains have started, I have been nothing but ill, . . . I shall soon pack up and go or I fear I will die here . . . I shall say good-bye to Asia forever."[1]

Rock left Kalimpong when he was 67 years old, in full command of his intellectual powers. He lived for another eleven years, resigned to a peripatetic life, first in Europe, then in the State of Washington, and finally in Hawaii; but always restless, always traveling, the perpetual emigrant, nowhere at home. Perhaps having left one country in his youth, he had left all countries; China, his adopted home, had spurned him.

Money was a problem for him, but Merrill's advice paid off and he was able to sell his library for $25,000 to the University of Washington's Far East and Russian Institute where, also, he was appointed as a permanent Honorary Research Associate and continued his studies for a time. In the late 1950s he went back to Hawaii and, during the last years of his life, divided his time between finishing his Nakhi dictionary and Hawaiian botany, to which he returned with enthusiasm and energy. Loy and Lester Marks, whom he had once guided around Peking, opened their home to him, and he lived as their guest for the last five years of his life in a style he never could have afforded on his own, released from financial worries for his scholarly interests and travel. He made several trips to Europe, one to South America in 1959, which he complained about but actually rather enjoyed, and finally to Asia with the Marks in 1961, but only to Japan, Hong Kong, and India, not to his precious China.

Rock was in Europe in 1962 classifying Nakhi manuscripts and making final arrangements for the printing of his dictionary which he had, at last, completed. He returned to Honolulu and died of heart failure on December 5, 1962, in the Marks home in upper Nuuanu Valley. His body lies in the Marks family plot in Hawaii, half way between West and East.

ACKNOWLEDGMENTS

I am indebted to the American Philosophical Society, the Ella Lyman Cabot Trust, and the National Geographic Society for financial assistance during the two and a half years I have worked on the manuscript of this book. I am also grateful to the Arnold Arboretum of Harvard University, the members of its staff, and especially its director, Dr. Richard A. Howard, for the loan of materials, for encouragement, and for assistance. Mr. M. V. Mathew, Librarian, and other members of the staff of the Royal Botanic Garden, Edinburgh, extended themselves most generously to facilitate my research on Rock's diaries. Mr. Glen Baxter, Associate Director of the Harvard-Yenching Institute, kindly rescued archival material for me from the bowels of the Institute.

For about two years Mr. Paul Weissich, Director of the Honolulu Botanic Gardens and executor of Joseph Rock's will, provided documents and answered bizarre questions with unfailing promptness and good humor; he was also very kind to me during my visit to Hawaii. Mrs. Loy Marks opened her home in Honolulu to me, let me snoop into her extensive collection of Rockiana, and provided me with many reminiscences of Rock, or "Pohaku," as she remembered him fondly. Mrs. Marks is an avid amateur of Hawaiian botany and is conscious of the significance of the work Rock did in that area; she expressed her disappointment when she learned that I planned to concentrate on his experiences in China. I hope, with her, that someone will one day publish a detailed study of the botanical contributions of this most extraordinary man. Mrs. Marks was also invaluable in giving me the names of persons in Hawaii who could assist me, and I spoke with several people. Among them I am especially grateful to Mr. E. H. Bryan and Dr. Harold St. John of the Bernice P. Bishop Museum, and to Mr. Horace Clay for their time and contribution.

Rock's nephew, Mr. Robert Koc, related family history which I could not have found anywhere else and introduced me to his brother, Hans, who filled me in on further details. Because my stay in Vienna was necessarily brief, Mr. Koc put himself at my disposal and talked for many hours about his uncle and the Rock family; he continued to be a most valued correspondent.

I interviewed many, many people, but I must single out the late Mr. Henry Corra, of Hong Kong, who charmed me during two visits to Boston, answered queries, and sent books which he thought might be of interest; Mr. Paul Fang, a Nakhi born in Likiang, Yunnan Province, where Rock lived for so many years; Mr. Terris Moore, who was with the Sikong Expedition which made the first ascent of Minya Konka; and Sir George Taylor, formerly Director of the Royal Botanic Gardens, Kew, England, who is the only person I ever met who called Rock "Joe." Among my several correspondents, I especially wish to thank Mr. W. L. Bond and Dr. Egbert Walker for taking time to write down their memories of Rock for me.

Mr. Paul Meyer, formerly an American Consul in Yunnanfu, China, invited me for lunch at his home in Connecticut—his wife, Harriet, makes a delicious cheese soufflee—and shared his knowledge of Rock and China. Both Mr. and Mrs. Meyer read and commented upon my manuscript, and I am sincerely grateful for their assistance.

I wish to express my special appreciation to two more people: first, the late Mr. Edgar Snow who, though ill (I did not yet know the state of his health) and burdened with work, took the time to answer my questions and encourage this project. I had always been an admirer of his and I prized his interest highly.

Second, Mr. Walter Muir Whitehill who took it upon himself to respond to my pleas for help, guided me in the proper direction, and cheered me when I most needed cheering. I think it is safe to say that without his moral support I might never have finished the book. Since I cannot repay him in kind, I only hope that I can do as much for someone else someday.

I have often been asked if I have been to China. There is a story behind this. After I had been at work on the manuscript for a while, I had a sudden urge to follow Rock's tracks through the mountains and river gorges of the western provinces. The doors to China were closed tight, however. Well, I thought, this can't go on forever. Sooner or later they will let an American in, and I bet it will be someone improbable. Why not me? Unduly optimistic, I wrote Peking to request a visa. The day after I dropped my letter in the mailbox, I heard that the U. S. table tennis team had been invited to China. I had to satisfy myself with the knowledge that my theory was correct; I wrote several more letters, but I never got to China. As far as I know, *no* Westerner has been permitted to travel recently in the Tibetan borderland area which Rock knew so well, but I sincerely hope that this part of the world will become accessible again soon.

S.B.S.

Boston, Mass.
November 1973

BIBLIOGRAPHY
AND OTHER SOURCES

The raw materials for this book were the unpublished diaries, letters, and a manuscript of Joseph Rock. According to Rock's will, the diaries are in the possession of the library of the Royal Botanic Garden, Edinburgh, Scotland; another set of journals was bequeathed to the *Deutsche Morgenlandische Gesellschaft* in Marburg, Germany, but this group contains only notes related to Rock's study of the Nakhi language and culture. The Arnold Arboretum of Harvard University, the Harvard-Yenching Institute, and the National Geographic Society all have large files of Rock's correspondence; smaller collections exist at the University of Hawaii, the Royal Botanic Garden, Edinburgh; the Royal Botanic Gardens, Kew (England); the Museum of Comparative Zoology, Harvard; the Hawaiian Sugar Planters' Association, Honolulu; and various other locations indicated in the notes as well as in the hands of private individuals. The University of Hawaii also has microfilm of some of the diaries which are in Edinburgh. The unpublished manuscript—draft of a book about adventures—was never completed and was inherited by Rock's nephew, Robert Koc, from whom it was purchased. It now belongs to the Arnold Arboretum.

I have listed only those published writings of Rock which are pertinent to this text. A complete bibliography of his works can be found in the *Hawaiian Botanical Society Newsletter,* January 1963, p. 10–13 (reprinted in *Taxon,* Vol. XII, No. 3, Apr. 1963, p. 98–102). This bibliography was compiled by Alvin K. Chock, E. H. Bryan, Jr., and Loy Marks. Mrs. Marks, an avid bibliophile and friend of Rock's, undoubtedly possesses the most complete collection of his published works.

BOOKS AND ARTICLES

Alley, Rewi, *China's Hinterland in the Leap Foreward,* Peking, New World Press, 1961.

Andrews, Roy Chapman, and Y. B. Andrews, *Camps and Trails in China,* New York, D. Appleton and Co., 1918.

Bretschneider, Emil, *European Botanical Discoveries in China,* London, Sampson, Low, Marston and Co., 1898.

Bryan, E. H., "An Anecdote Concerning Joseph F. Rock," *Hawaiian Botanical Society Newsletter* II (1), January 1963, p. 16–17.

Buxton, L. H. D., *China; The Land and the People,* Oxford, The Clarendon Press, 1929.

Carlquist, Sherwin, *Hawaii: A Natural History,* Garden City, New York, Natural History Press, 1970.

Chiang Kai-shek, *Soviet Russia in China,* New York, Farrar, Straus and Cudahy, 1957.

Chock, Alvin K., "J. F. Rock, 1884–1962," *Hawaiian Botanical Society Newsletter,* II: 1, January 1963, p. 1–13. (Reprinted in *Taxon,* XIII:3, April 1963.)

Clubb, O. Edmund, *Twentieth Century China,* New York, Columbia University Press, 1965. (paper.)

Cordier, G., *La Province du Yunnan,* Hanoi, Le-Van-Tan, 1928.

Cowan, J. M., *The Journeys and Plant Introductions of George Forrest,* London, Oxford University Press for the Royal Horticultural Society, 1952.

Cox, E. H. M., *Plant Hunting in China,* London, Collins, 1945.

Cressey, George B., *China's Geographic Foundations,* New York, McGraw-Hill, 1934.

David-Neel, Alexander, *A l'Oeust Barbare de la Vaste China,* Paris, Librairie Plon, 1947.

Davies, H. R., *Yunnan; the Link between India and the Yangtze,*

Cambridge, England, University Press, 1909. (Reprinted by Ch'eng Wen Publishing Co., Taipei, 1970.)

Davies, John Paton Jr., *Dragon by the Tail*, New York, Norton, 1972.

Duncan, Marion H., *The Mountain of Silver Snow*, Cincinnati, Powell & White, 1929.

———, *The Yangtze and the Yak*, Alexandria, Va., privately printed, 1952.

Edgar, James H., in *Journal of the West China Border Research Society*, Vol. I, 1922–23, p. 59.

Fairbank, John K., *The United States and China*, Cambridge, Mass., Harvard University Press, 1958. (paper, 1971)

Farrar, Reginald, "My Second Year's Journey on the Tibetan Border of Kansu," *Geographical Journal*, LI, No. 6, June 1918, p. 341–359.

Fitzgerald, C. P., *The Tower of Five Glories; A Study of the Min Chia of Ta Li, Yunnan*, London, The Cresset Press, 1941.

Fleming, Peter, *News from Tartary*, New York, Charles Scribner's Sons, 1936.

Fox, Helen M., (editor and translator), *Abbé David's Diary*, Cambridge, Mass., Harvard University Press, 1949.

Goullart, Peter, *Forgotten Kingdom*, London, John Murray, 1955. (paper).

High, Stanley, "China Astir Against the Foreigner," *Asia*, XXV, No. 8, August 1925, p. 652–655, 705–707.

Hosie, Alexander, *Three Years in Western China; A Narrative of Three Journeys in Ssu-ch'uan, Kuei-chow, and Yun-nan*, London, George Philip & Son, 1890.

Hsu, Immanuel C. Y., *The Rise of Modern China*, New York, Oxford University Press, 1970.

Hu, Chang-tu, *China: Its People, Its Society, Its Culture*, New Haven, Human Relations Area File, Inc., 1960.

Jacoby, Annalee, and Theodore H. White, *Thunder out of China*, New York, William Sloane Associates, Inc., 1946.

Karamisheff, W., *Mongolia and Western China*, Tientsin, La Librairie Française, 1925.

Kingdon Ward, Frank, *From China to Hkamti Long*, London, Edward Arnold & Co., 1924.

———, *The Romance of Plant Hunting*, London, Edward Arnold & Co., 1924.

Lattimore, Owen, *Inner Asian Frontiers of China*, New York, American Geographical Society (Research Series No. 21), 1940.

Leong, Y. K., and L. K. Tao, *Village and Town Life in China*, London, George Allen & Unwin, Ltd., 1915.

Maurer, Herrymon, *The End Is Not Yet*, New York, Robert M. McBride & Co., 1941.

d'Ollone, H. M. G., *Les Derniers Barbares*, Paris, Pierre Lafitte & Cie., 1911.

Peffer, Nathaniel, "The Uniqueness of Missionaries," *Asia*, XXIV, No. 5, May 1924, p. 353–357.

———, "China before the Annual Armistice," *Asia*, XXIV, No. 12, Dec. 1924, p. 996.

Riley, J. H., "Three New Birds from Western China," *Biological Society of Washington*, vol. 30, January 27, 1925, p. 9–12.

Rock, Joseph F., *The Ancient Na-khi Kingdom of Southwest China*, (2 vols.), Harvard-Yenching Monograph Series VIII, Cambridge, Harvard University Press, 1947.

———, "Banishing the Devil of Disease among the Nashi: Weird Ceremonies Performed by an Aboriginal Tribe in the Heart of Yunnan Province, China," *National Geographic Magazine*, XLVI, 1924, p. 473–499.

———, "Butter as a Medium of Religious Art," *Illustrated London News*, 175 (4721), 1929, p. 636–639.

———, "Choni, the Place of Strange Festivals," *Illustrated London News*, 175 (4718), p. 520, 1929.

———, "Expedition to Tibet of the National Geographic Society," *Science*, new series, 58, p. 460, 1923.

———, "Experiences of a Lone Geographer: An American Agricultural Explorer Makes His Way through Brigand-Infested Central China En Route to the Amne Machin Range, Tibet," *National Geographic Magazine*, XLVIII, 1925, p. 331–347.

———, "Glories of the Minya Konka: Magnificent Snow Peaks of the China-Tibetan Border are Photographed at Close Range by a National Geographic Society Expedition," *National Geographic Magazine*, LVIII, 1930, p. 385–437.

———, "Hunting the *Chaulmoogra* Tree," *National Geographic. Magazine*, XLI, 1922, p. 243–276.

———, *The Indigenous Trees of the Hawaiian Islands*, Honolulu, 1913.

———,"Konka Risumgongba, Holy Mountain of the Outlaws," *National Geographic Magazine,* LX, 1931, p. 1–65.

———, "Land of the Tebbus," *Geographical Journal,* vol. 81, 1933, p. 108–127.

———, "Land of the Yellow Lama: National Geographic Society Explorer Visits the Strange Kingdom of Muli, Beyond the Likiang Snow Range of Yunnan Province, China," *National Geographic Magazine,* XLVII, 1925, p. 447–491.

———, "Life Among the Lamas of Choni: Describing the Mystery Plays and Butter Festival in the Monastery of an Almost Unknown Tibetan Principality in Kansu Province, China," *National Geographic Magazine,* LIV, 1928, p. 569–619.

———, *The Ornamental Trees of Hawaii,* Honolulu, 1917.

———, "Seeking the Mountains of Mystery: An Expedition on the China-Tibet Frontier to the Unexplored Amnyi Machen Range, One of Whose Peaks Rivals Everest," *National Geographic Magazine,* LVII, 1930, p. 131–185.

———, "The Story of the Flood in the Literature of the Mo-so (Nakhi) Tribe," *Journal of the West China Border Research Society,* VII, 1935, p. 64–80.

———, "Sungmas, the Living Oracles of the Tibetan Church," *National Geographic Magazine,* LXVIII, 1935, p. 475–486.

———, "Through the Great River Trenches of Asia: National Geographic Society Explorer Follows the Yangtze, Mekong, and Salwin Through Mighty Gorges, . . ." *National Geographic Magazine,* L. 1926, p. 133–186.

———, "The Voyage of the *Luka* to Palmyra Island," *Atlantic Monthly,* vol. 144, no. 9, 1929, p. 360–366.

Roosevelt, Kermit, and Theodore Roosevelt, *Trailing the Giant Panda,* New York, Charles Scribner's Sons, 1929.

Rue, John E., *Mae Tse-tung in Opposition, 1927–1935,* Stanford, Calif., Stanford University Press (for the Hoover Institution on War, Revolution and Peace), 1966.

Salisbury, Harrison E., *Orbit of China,* Harper & Row, New York, 1967.

Sheridan, James E., *Chinese Warlord; the Career of Feng Yu-hsiang,* Stanford, Calif., Stanford University Press, 1966.

Shih, Chan-chun, *The Conquest of Minya Konka,* Peking, Foreign Language Press, 1959.

Snow, Edgar, *The Battle for Asia*, New York, Random House, 1941.

——, *Journey to the Beginning*, New York, Random House, 1958.

——, *The Other Side of the River*, New York, Random House, 1961 (Reissued in 1971 as *Red China Today* by Vintage Books, in paper.)

——, *Red Star over China*, New York, Random House, 1938. (Reissued in 1961 by Grove Press Inc., in paper.)

Sutton, S. B., *Charles Sprague Sargent and the Arnold Arboretum*, Cambridge, Mass., Harvard University Press, 1971.

Teichman, Eric, *Travels in North-west China*, Cambridge, England, University Press, 1921.

Toland, John, *The Rising Sun* (2 vols.), New York, Random House, 1970.

Tuchman, Barbara W., *Stilwell and the American Experience in China, 1911–45*, New York, The Macmillan Co., 1970.

U.S. Department of State, *Foreign Relations of the United States*, annual series, Washington, Government Printing Office, various dates. Cited in Notes as USFR.

Wilson, Dick, *The Long March*, New York, The Viking Press, 1971.

Wilson, Ernest H., *Aristocrats of the Garden*, Boston, Stratford, 1926.

Younghusband, Sir Francis (compiler), *Peking to Lhasa;* The Narrative of the Journeys in the Chinese Empire Made by the Late Brigadier-General George Pereira, London, Constable and Co., Ltd., 1925.

NOTES

Unless otherwise indicated, direct quotations from Rock are drawn from his personal journals.

Abbreviations after citations of letters indicate the location of that particular letter and are as follows: AA = Arnold Arboretum of Harvard University, Jamaica Plain, Mass.; NGS = National Geographic Society, Washington, D. C.; HY = Harvard-Yenching Institute, Cambridge, Mass. NGM is National Geographic Magazine.

I

1. Rock, "Land of the Yellow Lama"; NG, XLVII, 1925, p. 451
2. Wilson, as quoted in Sutton, *Charles Sprague Sargent*, p. 213
3. I am indebted to Ms. Eva Hesse, a Pound scholar, for pointing out the poet's familiarity with Rock's scholarship. As far as I can ascertain, there was no correspondence between Pound and Rock; I doubt that Rock even knew of Pound's interest in his work or read a single line of his poetry. The quotation is from Pound's *Canto CXII*.
4. Goullart, *Forgotten Kingdom*, p. 18.
5. Rock, "Land of the Yellow Lama . . . " p. 451.
6. Père Jean Marie Delavay made the first extensive botanical collection in the Likiang vicinity in the 1880s and '90s; see Bretschneider. p. 874–911.
7. Edouard Chavannes, *Documents historiques et géographiques relatifs à Li-Kiang*, published what Rock believed to be an error-riddled translation of a Chinese chronicle dealing with Nakhi history in 1912. Jacques Bacot, who traveled in Likiang in 1907 and 1909, saw the same chronicle and had the portraits of Nakhi kings copied; in his published account, *Les Mo-so* (1913), Rock claims he repeated many of Chavannes' mistakes. See Rock, *The Ancient Na-khi Kingdom*, vol. 1, p. 67–69 for further discussion.
8. Augustine Henry, the Irish Customs officer-botanist, relieved the monotony of his life in Szemao by compiling a Lolo dictionary which was never published.
9. Rock to David Fairchild, April 22, 1923, copy NGS.

II

1. Chock, "J. Rock, 1884–1962," p. 2. Mr. Otto Degener supplied a few additional details.
2. This is the version which Rock told his nephew, Robert Koc.

3. Note from a field book at the Bernice P. Bishop Museum, Honolulu, May 30, 1910.

4. Mr. Anthony Halasz kindly double-cheched this information for me with the University of Vienna in June 1971.

5. Bryan, "An Anecdote Concerning Joseph F. Rock," p. 16, and personal communication.

6. For the complete list, see the bibliography which accompanies Chock's remembrance of Rock.

7. For those unfamiliar with botany or the Hawaiian islands it is pertinent to comment lest they remain unimpressed by Rock's accomplishments. Hawaii is a little larger in land area than Connecticut. The islands are volcanic in origin. Of the principal islands, Kauai, the oldest, is probably between 5.6 and 3.8 million years old; the most recent, Hawaii, appears to have been formed largely within the last million years and claims two active volcanoes, Mauna Loa and Kilauea. The terrain on all the islands is dramatic, varying from one loction to the next. The islands include several ecological zones within their modest area, and these zones support distinct floras. Rock divided Hawaii into neat botanical regions, and no one has managed to improve his outline. Besides a wide range of vegetative types to capture the imagination, the Hawaiian flora, by virtue of geographic isolation, shows the highest percentage of endemism (i.e., of forms which occur *only* in Hawaii), at both genus and species level, of any area in the world. All these factors provide fun for naturalists and, since the flora is richer than the fauna, particularly for botanists. Interested readers are referred to Sherwin Calquist's *Hawaii: A Natural History.*

Horace Mann, Jr., a protegee of Harvard's venerable Asa Gray, botanized in Hawaii in the 1860s and published his *Enumeration of Hawaiian Plants* in 1867. But the only botanist of real consequence before Rock in Hawaii was Dr. William Hillebrand (1821–1866), a German-born physician who arrived in the islands in 1851. Like Rock he had tuberculosis and, also like Rock, he recovered in the Hawaiian sunshine. Hillebrand distinguished himself in several ways. As a member of the Privy Council of King Kamehameha V, he concerned himself with the economic growth of the islands; recognizing the shortage of agricultural workers, he encouraged the importation of labor from China, India, Malaya, and of Portuguese from Madeira, thereby influencing the texture of human life as well as the character of the landscape. He introduced plants and birds to Hawaii, taking particular interest in plant species of economic or ornamental value. He made the first significant collection of Hawaiian plants, and his *Flora of the Hawaiian Islands,* published two years after his death, constituted the first important literature on the subject. Next in line was Rock's own *Indigenous Trees,* a publication sponsored by a group of wealthy patrons whom Rock had interested in his work. Their generosity permitted the inclusion of 218 plates (Hillebrand's *Flora* had only one illustration), mostly the work of Rock's camera which he handled with professional expertness.

8. Rock, "The Voyage of the Luka to Palmyra Island," *Atlantic Monthly*, 144 (9) 1929, p. 360, 366.

9. Bryan, "An Anecdote Concerning Joseph F. Rock," p. 16.

10. This entire episode is related by Rock in his unpublished manuscript.

11. Rock, *"Hunting the Chaulmoogra Tree,"* p. 266.

12. Cox, *Plant Hunting in China*, p. 80–81.

13. Archibald Little, introduction to Hosie, *Three Years in Western China*, p. xvi.

14. Bretschneider's *European Botanical Discoveries in China* is the standard reference work for all botanical exploring in the Far East before 1900.

III

1. Lattimore, *Inner Asian Frontiers of China*, p. 22.

2. Fox, *Abbe David's Diary*, p. 253.

3. Wilson, *Aristocrats of the Garden*, p. 275–294.

4. See Cordier, *La Province de Yunnan*, p. 530–556.

5. USFR, 1927, vol. II, p. 29, MacMurray to the Secretary of State.

6. *ibid.*, p. 306–307, MacMurray to the Secretary of State.

7. Cordier, *La Province de Yunnan*, p. 205–210.

8. Fitzgerald, *The Tower of Five Glories*, p. 28.

9. Tuchman, *Stilwell . . .* ,p. 364.

10. Cressey, *China's Geographic Foundations*, p. 316.

11. USFR, 1930, vol. II, p. 7. Political Report.

12. Peffer, "China before the Annual Armistice," p. 996.

13. USFR, 1925, vol. I, p. 799, MacMurray to the Secretary of State.

14. USFR, 1926, vol. I, p. 999, MacMurray to the Secretary of State.

15. Peffer, "The Uniqueness of Missionaries," p. 357.

16. Rock to C. S. Sargent, September 12, 1925, AA.

17. Rock to David Fairchild, November 5, 1923, copy NGS.

IV

1. Riley, "Three New Birds from Western China."

2. Sutton, *Charles Sprague Sargent.*

3. Pereira, as quoted in Younghusband, *Peking to Lhasa*, p. 117.

4. Rock to C. S. Sargent, undated, AA.

5. Rock to C. S. Sargent, November 5, 1924, AA.

6. diary, December 13, 1924.
7. Henry Corra, personal communication.
8. Duncan, *The Yangtze and the Yak,* p. 37.
9. Rock to C. S. Sargent, January 27, 1925, AA.
10. Pereira, as quoted in Younghusband, *Peking to Lhasa,* p. 266.
11. Rock to C. S. Sargent, February 18, 1925, AA.
12. diary, March 14, 1925.
13. Rock, unpublished manuscript.
14. Rock to C. S. Sargent, February 21, 1925, AA.
15. Teichman, *Travels in North-west China,* p. 148.
16. Rock, unpublished manuscript.
17. Rock to C. S. Sargent, May 10, 1925, AA.
18. Rock to C. S. Sargent, May 26, 1925, AA.
19. Rock to C. S. Sargent, May 26, 1925, AA.
20. Lu Hung-t'ao to Rock, copy included with Rock to C. S. Sargent, June 5, 1925, AA.
21. Lu Hung-t'ao to Rock, June 9, 1925, copy AA.
22. Rock to F. V. Coville, September 18, 1925, copy NGS.

v

1. Mr. Paul Weissich, executor of Rock's will, was kind enough to loan me the address book along with letters which Rock had received near the end of his life.
2. Rock's nephew Robert Koc, vaguely remembered his uncle showing him a picture of a person Robert took to be a girlfriend in Hawaii; no one there recalls Rock ever being romantically linked. Dr. Harold St. John said he once met a European lady in the Berlin herbarium who claimed she expected to become Mrs. Rock, but there is no further confirmation of that story either.
3. Snow, *Journey to the Beginning,* p. 57, 59.
4. Farrar, *Tibetan Border of Kansu.*

VI

1. diary, February 25, 1926.
2. Rock to Ralph Graves, February 27, 1926, NGS.
3. That Feng initially acted within a legal framework was a freakish accident of history. He had held the post of *Tupan,* or Commissioner, of Northwestern Defense since May 1923, but did not seek to avail himself of its power until he had been forced from Peking in January 1926. He resigned the *tupan*-ship later, under further pressure from eastern warlords. See Sheridan, *Chinese Warlord,* p. 187.

4. Sheridan, *Chinese Warlord*, p. 194–195.
5. diary, May 5, 1926.
6. Rock to C. S. Sargent, June 8, 1926, AA.
7. d'Ollone, *Les Derniers Barbares*, p. 240–241.
8. Rock, "Seeking the Mountain of Mystery" p. 160.
9. E. F. Stanton to Rock, June 8, 1926, AA.
10. Sheridan, *Chinese Warlord*, p. 197.

VII

1. Fairbank, in *The United States and China,* gives a capsule description of Chinese bureaucracy and its attendant evils, p. 96–105.
2. Rock to Ralph Graves, March 22, 1928, NGS.
3. Kingdon Ward, *The Romance of Plant Hunting*, p. 140.
4. Rock to Franklin L. Fisher, March 26, 1935, NGS.
5. Rock, *The Ancient Na-khi Kingdom* . . . vol. II, p. 355–356. Rock observes a conflicting record which indicates that Chienso and Tsoso *T'ussus* were appointed as early as the Wan-li period, 1573–1620.
6. Rock, *The Ancient Na-khi Kingdom* . . . vol. II, p. 402.
7. Cressey, *China's Geographic Foundations,* p. 391.
8. Rock, *The Ancient Na-khi Kingdom* . . . vol. II, p. 355.
9. Rock, *The Ancient Na-khi Kingdom* . . . vol. II, p. 408.

VIII

1. Rock to Ralph Graves, May 24, 1927, NGS.
2. Memorandum, O. LaGorce to F. V. Coville, September 27, 1927, copy NGS.
3. Memorandum, A. Bumstead to O. LaGorce, September 22, 1927, copy NGS.
4. Memorandum, Ralph Graves to W. Simpich, December 31, 1928, NGS.
5. Nelson T. Johnson to Secretary of Agriculture, September 27, 1927, NGS.
6. Rock to Ralph Graves, November 25, 1927, NGS.
7. Rock to Graves, September 22, 1928, NGS.
8. diary, March 29, 1928.
9. Rock to G. Grosvenor, May 1, 1928, NGS.
10. Text of telegram repeated in Rock to Graves, September 22, 1928, NGS.
11. Rock to Graves, September 22, 1928, NGS.
12. diary, January 4, 1929.
13. Rock, unpublished manuscript.

14. Jack Young, highly regarded by the Roosevelts, acted as interpreter and guide for the 1932 Sikong Expedition which achieved the first climb of Minya Konka. He served with the Chinese war government and the U.S. Army druing World War II. After the war he became an ADC to General Marshall and, in that capacity, acted as an interpreter between Marshall and Mao Tse-tung. He retired as a Brigadier-General of the U.S. Army and lives in the Midwest.

15. Rock, unpublished manuscript.

16. The Roosevelts good-naturedly admitted to some lack of forethought in *Trailing the Giant Panda,* their published version of the expedition, and one has to believe, with Rock, that they were lucky to have survived their journey with as few mishaps as they had. But throughout their narrative one senses adventure without unnecessary tension, and a kind of joy that is conspicuously absent from Rock's writings. One longs for a happy medium between the two attitudes.

17. diary, February 7, 1929.

18. Rock to Graves, March 20, 1929, NGS.

19. Rock to Graves, October 20, 1929, NGS. I spoke with Young about the article in question, and he had no specific recollection of it, having written a number of articles in those days. However, he assured me that he never intended any harm to Rock and had only the highest respect for his work. He remembered that after the war, when he was in the service of the Nationalist government which became interested in the mineral resources of the Tibetan borderland, an official asked him 'and what about Rock? What's he doing out there?' Young replied that if anyone had opened up that part of China, both to the world and to the Chinese themselves, it had been Rock. When the official still seemed concerned, Young pointed out that Rock was not going to make off with either the region or its minerals.

20. Shih, *The Conquest of Minya Konka,* p. 50. The priest was probably Dominicus Parennin; cf. Bretschneider, p. 19–20.

21. Edgar, in *Journal of the West China Border Research Society,* p. 59.

22. Rock to NGS, February 27, 1930, NGS.

23. Press Release from NGS, undated, ca. March 1930, NGS.

24. diary, August 7, 1929.

25. diary, August 24, 1929.

26. Stanley Hornbeck to William Simpich, December 7, 1929, NGS.

27. Rock to Graves, October 20, 1929, NGS.

28. Memorandum, William Simpich to O. LaGorce, May 23, 1930, NGS.

29. diary, November 29, 1929.

30. diary, December 19, 1929.

31. Rock to Graves, December 30, 1929, NGS.

32. Rock to Graves, September 18, 1929, NGS.

33. diary, December 23, 1929.

IX

1. Rock to Graves, October 6, 1930, NGS.
2. Rock to Alfred Rehder, April 7, 1930, AA.
3. There is no doubt that in terms of numbers of ornamental species introduced into American gardens from western China, Wilson was the botanical collector *par excellence,* but this can be explained as much by circumstance as by skill. He was the first large scale collector to work an area with an abundant flora, in Hupeh and Szechwan. Later botanists, exploring to the north and west of Wilson country, found fewer species or often the same ones Wilson had already introduced—a situation which made them resentful of Wilson's good luck in having been first in the best localities. Frank Meyer and William Purdom both expressed twinges of professional jealousy on at least one occasion. Yet these men and Rock performed a great service to science with their studies of remote floras. Wilson was primarily interested in garden plants and underrated the scientific value of Rock's work.
4. Rock to Graves, October 6, 1930, NGS.
5. USFER, 1930, vol. II, p. 46. Minister in China to Secretary of State. Political Report No. 38.
6. Rock to Graves, October 6, 1930, NGS.
7. Snow, *Journey to the Beginning,* p. 56.
8. Rock to Oakes Ames, March 11, 1931, AA.
9. Rock, *The Ancient Na-khi Kingdom* . . . vol. I, p. 265.
10. Rock to Graves, April 8, 1931, NGS.
11. Snow, *Journey to the Beginning,* p. 55.
12. Fairbank, *The United States and China,* p. 294.
13. Tuchman, *Stilwell* . . ., p. 96.
14. USFR, 1930, vol. II, p. 57, Monthly Report for October, 1930.
15. ibid., p. 208, Secretary of State to Rev. J. J. Burke.
16. Davies, *Yunnan..........,* p. 163–164.
17. Clubb, *Twentieth Century China,* p. 181–182, 187.
18. Rock to Graves, November 26, 1931, NGS.

X

1. Snow, *Red Star Over China,* p. 199.
2. See Wilson, *The Long March,* p. 134–150.
3. Rock to Franklin Fisher, May 23, 1936, NGS.
4. Diary, April 16, 1936. The dates Rock gives do not coincide with Wilson, p. 240, who indicates April 23 as the day of the Second Front Army approached Yunnanfu; Rock's dates seem more precise in this instance.

5. Snow, *Red Star Over China*, p. 62.
6. Rock to Franklin Fisher, May 23, 1936, NGS.
7. Rock to Fisher, May 23, 1936, NGS.
8. Tuchman, *Stilwell* . . . , p. 145.

XI

1. USFR, 1937, vol. IV, p. 376, Commander in Charge of the U.S. Asiatic Fleet to Ch. Naval Operations.
2. *Honolulu Star-Bulletin*, October 25, 1938.
3. *Honolulu Advertiser*, August 28, 1940. Crawford was forced into premature retirement three weeks after the blow-up with Rock. An article in the *Hawaii Sentinel*, September 19, 1940, hinted at Crawford's Nazi sympathies and mentioned the Rock incident in connection with the resignation.
4. *Honolulu Star-Bulletin*, August 27, 1940.
5. Rock, unpublished manuscript.
6. Jacoby and White, *Thunder out of China*, p. 140.
7. W. L. Bond, personal communication.

XII

1. Owing to 80% duplication, Harvard-Yenching ultimately declined Rock's library. He sold the collection to the University of Washington for $25,000 in the early 1950s.
2. Carl T. Keller, to Elisseeff, April 27, 1945, HY.
3. See correspondence in Harvard-Yenching files for Rock for 1948.
4. *Honolulu Advertiser*, October 1, 1946.
5. Rock to Wallace B. Donham, January 3, 1947, HY.
6. ibid.
7. Rock to E. D. Merrill, January 29, 1947, AA.
8. Rock to Merrill, January 9, 1947, AA.
9. Rock to Elisseeff, July 17, 1947, HY.
10. Rock to Merrill, April 26, 1947, AA.
11. It was actually published in 1952 by the Istituto Italiano per il Medio ed Estremo Oriente.
12. Fairbank, *The United States and China*, p. 312.
13. Hsu, *The Rise of Modern China*, p. 738.
14. Rock to Elisseeff, December 19, 1947, HY.
15. Rock to Merrill, January 28, 1948, AA.
16. Rock to Egbert Walker, November 1, 1948, AA.

17. Rock to Elisseeff. April 5, 1949, HY; also, Goullart, *Forgotten Kingdom*, p. 199–206.

18. Goullart, *Forgotten Kingdom*, p. 213.

19. Snow, *Journey to the Beginning*, p. 197–207, gives information on IN-DUSCO.

20. Rock to Merrill, August 24, 1949, AA.

EPILOGUE:

1. Rock to Andrew Tse, June 10, 1951, courtesy of Mr. Tse.

INDEX